An Aether expressing Fibonacci and Lucas Sequences joined at the hip: a thought experiment

Table of Contents

An Aether expressing Fibonacci and Lucas Sequences joined at the hip: a thought experiment

An Aether expressing Fibonacci and Lucas Sequences joined at the hip: a thought experiment

3

An Aether expressing Fibonacci and Lucas Sequences joined at the hip: a thought experiment

An Aether expressing Fibonacci and Lucas Sequences joined at the hip: a thought experiment

An Aether expressing Fibonacci and Lucas Sequences joined at the hip: a thought experiment

An Aether expressing Fibonacci and Lucas Sequences joined at the hip: a thought experiment

7

An Aether expressing Fibonacci and Lucas Sequences joined at the hip: a thought experiment

8

An Aether expressing Fibonacci and Lucas Sequences joined at the hip: a thought experiment

An Aether expressing Fibonacci and Lucas Sequences joined at the hip: a thought experiment

An Aether expressing Fibonacci and Lucas Sequences joined at the hip: a thought experiment

11

An Aether expressing Fibonacci and Lucas Sequences joined at the hip: a thought experiment

An Aether expressing Fibonacci and Lucas Sequences joined at the hip: a thought experiment

An Aether expressing Fibonacci and Lucas Sequences joined at the hip: a thought experiment

An Aether expressing Fibonacci and Lucas Sequences joined at the hip: a thought experiment

An Aether expressing Fibonacci and Lucas Sequences joined at the hip: a thought experiment

16

An Aether expressing Fibonacci and Lucas Sequences joined at the hip: a thought experiment

Dedication:

To my mother Anne, my sister Virginia, and my father Cedric.

An Aether expressing Fibonacci and Lucas Sequences joined at the hip: a thought experiment

Introduction

As is well known, the Fibonacci sequence is named for the mathematician Leonardo Fibonacci, also known as Leonardo Pisano, who lived in the 12th and 13th centuries. Well-travelled, he left his birth city of Pisa and studying mathematical calculation under an Arab master in Algeria, Egypt, Syria, Greece, Sicily and Provence. No biography would be complete of Leonardo Fibonacci, regardless of how short, without also mentioning his two most famous works "Liber abaci" (1202; "Book of the abacus") and "Liber quadratorum" (1225; "Book of the square numbers"). It was through the former book that Fibonacci first introduced the use Hindu-Arabic numerals into European consciousness for common arithmetical operations. It the former book was also the first to pose the problem [1]:

> A certain man put a pair of rabbits in a place surrounded on all sides by a wall. How many pairs of rabbits can be produced from that pair in a year if it is supposed that every month each pair begets a new pair which from the second month on becomes productive?

which results in the number sequence 1, 1, 2, 3, 5, 8, 13, 21, 34, 55 in which each number is the sum of the two proceedings numbers. It took another three centuries before French-born mathematician Albert Girard, in 1634, deduced an expression of the sequence of numbers through the formula:

- $U_{n+2} = U_{n+1} + U_n.$

And thus is the short biopic of the numbers of the Fibonacci sequence as we know them today.

An Aether expressing Fibonacci and Lucas Sequences joined at the hip: a thought experiment

The Lucas sequence is named for the French mathematician François Édouard Anatole Lucas [2] 1842–1891). As with the Fibonacci numbers, each Lucas number is defined to be the sum of its two immediately previous terms, thereby forming a Fibonacci integer sequence [3]. The two sequences differ only in terms of their initial conditions: whereas the Fibonacci sequence begins with

- $F(0) = 0$ and $F(1) = 1$

the Lucas sequence begins with

- $L(0) = 2$ and $L(1) = 1$.

It was also Lucas who first coined the term "Fibonacci sequence" in the 19th century, and Lucas numbers and Fibonacci numbers are now described as instances of Lucas sequences [4]. It was in the 19th century that Lucas and other scientists began to remark on the appearance of numbers related to the Fibonacci and Lucas sequences in nature. These included the spirals of sunflower heads, in pinecones, in the regular descent (genealogy) of the male bee, in the related logarithmic (equiangular) spiral in snail shells, in the arrangement of leaf buds on a stem, and in animal horns.

The purpose of this book however is not to document these expressions and arrangements in nature. It is rather the premise of this book that the appearance of Fibonacci and Lucas numbers are an epiphenomenon of a more fundamental sub-structure of the universe, in essence an Aether, which penetrates matter and space.

That is, it is the premise of this book that if the objects we see in the Universe can be considered to be formed from an omni-present Aether, this book is an exploration of an

An Aether expressing Fibonacci and Lucas Sequences joined at the hip: a thought experiment

example of an "Aether-based Universe" derived from values in each of the Fibonacci and Lucas Sequences. It is therefore intended in this book to investigate:

- how "values" for these numbers are generated

and

- how "values" (or Aether-States) based on the numbers may interact with one another

to define the properties of the universe as we experience it.

An Aether expressing Fibonacci and Lucas Sequences joined at the hip: a thought experiment

Disclaimer

- This is document is not to be considered an authorized textbook. The contents of this document should only be considered to be best guesses or food for thought.

- No experiments have been carried out by the author to either verify or falsify any aspect of the ideas put forward in this document. If any reader wishes to carry out experiments to "falsify" (- i.e. to put to the test) any aspect of the ideas in this document, they do so at their own risk.

- Prospective researchers, seeking inspiration for requests of a grant for their next PhD or Doctorate, are encouraged to investigate the claims in this book and at worst, prove me wrong.

An Aether expressing Fibonacci and Lucas Sequences joined at the hip: a thought experiment

Standard Generation of the Fibonacci Sequence

A familiar expression for generating numbers in the Fibonacci Sequence involves a function $F(n)$ involving ever increasing values of the variable "n".

Positive Advancing Sequence

As "n" advances forward, starting from initial values of "n" of "0" and "1" being assigned values of zero and one, the Sequence evolves:

- $F(0) = 0$

- $F(1) = 1$

- $F(2) = 1 \ [= 1 + 0]$

- $F(3) = 2 \ [= 1 + 1]$

- $F(4) = 3 \ [= 2 + 1]$

- $F(5) = 5 \ [= 3 + 2]$

Where each value is determined by recurrence relation

$$F(n) = F(n-1) + F(n-2) \qquad \textbf{[EQN(1)]}$$

What is sometimes overlooked is that a complementary Fibonacci Sequence can be built using a function $F(-n)$ involving negative values of "n" which are ever decreasing.

22

An Aether expressing Fibonacci and Lucas Sequences joined at the hip: a thought experiment

Negative Advancing Sequence

In this case, as the direction of "n" reverses backward, starting from initial values of "n" of "0" and "1" also being assigned values of zero and one, the Sequence evolves

- $F(0) = 0$

- $F(-1) = +1$

- $F(-2) = -1 \; [= 0 - 1]$

- $F(-3) = +2 \; [= 1 - (-1)]$

- $F(-4) = -3 \; [= -1 - 2]$

- $F(-5) = +5 \; [= 2 - (-3)]$

Where each value is determined by a second recurrence relation

$$F(-n) = F(-n+2) - F(-n+1) \qquad \textbf{[EQN(2)]}$$

Combining positive and negative advancing sequences

If we were to view the values obtained from EQN1 and EQN2 on a cartesian graph, "n" would form a number line extending in left and right directions from the origin of "0", with EQN(1) being used to produce results on the positive side of the origin and EQN(2) being used to produce results on the negative side. Neither equation being destined to meet again.

Such a graph assumes however that the ordinates achieved representing the values of the Fibonacci Sequence need only be achieved by the two recurrence equations

An Aether expressing Fibonacci and Lucas Sequences joined at the hip: a thought experiment

described. That is, that the values obtained are achieved through simple addition / subtraction of integer values in the Real number set.

There are however other ways to reproduce values of the Fibonacci Sequence involving complex values.

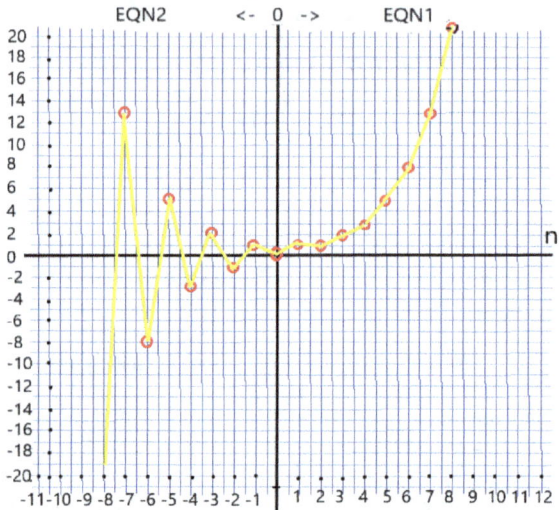

Graph of values of the Fibonacci Sequence obtained through equations EQN(1) and EQN(2) with "n" progressing in a positive direction (- >) and a negative direction (< -).

An Aether expressing Fibonacci and Lucas Sequences joined at the hip: a thought experiment

. As described, previous recursive functions for generating values of the Fibonacci Sequence involved the premise that either:

- $F(1) = 1$ and $F(2) = 1$

or

- $F(0) = 0$ and $F(1) = 1$

and from there the recursive equation of EQN(1) was enough to predict the subsequent values. This recursive equation however does not explain why relationships involving values of Fibonacci Sequences turn up in nature for example. The premise here is that the abundance of examples of Fibonacci Sequence values in nature, i.e., from ripples on sand dunes to arrangement of leaves and petals on the respective stems of a tree or the head of a flowering plant, is down to something much more fundamental.

Indeed, it is submitted that the behaviour of the natural world in terms of reflecting the behaviour of values of the Fibonacci Sequence and values of derivative sequences of the Fibonacci Sequence (for example the Lucas Sequence), come down to how we define fundamental values of "1", "0" and "2" in relation to one another.

25

An Aether expressing Fibonacci and Lucas Sequences joined at the hip: a thought experiment

An Aether-based Generation of the Fibonacci Sequence and the Lucas Sequence

The following is a description of a different expression for generating number values in the Fibonacci Sequence. This expression involves interleaved sequences of sums of squares and differences of squares.

Reactance in generating Fibonacci sequences.

Here the switching of polarities involves attributed to complex values **[5]** of a = b + ic where "a" and "b" are integers in the Real number set and "ic" is the imaginary component, where:

- "i" = square root of -1 (- i.e., $\sqrt{-1}$).

Furthermore, it is also the premise of this book that the nature of the proposed Aether is also considered to have characteristics giving rise to properties of rotation of bodies about an axis, revolution of bodies about one another, and direction of spin of rotation. It is thus proposed that another field of mathematics that may prove useful in determining the effects of an Aether is the use of quaternions **[6]**. These are an extension of complex numbers with the caveat that the values thereof comprise one Real number component and three imaginary components "i", "j" and "k". These imaginary components have the properties:

- $I^2 = j^2 = k^2 = ijk = -1$

An Aether expressing Fibonacci and Lucas Sequences joined at the hip: a thought experiment

Dualling numbers in closed circuits modular number systems

In an additional discussion on the Cassini Identity and the Catalan Identity regarding the generation of Fibonacci and Lucas numbers, it will be further explored how said numbers may be considered to have "dual number ε" qualities ascribed to them [7], where:

- $\varepsilon^2 = 0$.

In particular, the "dual number ε" qualities of these generated values are considered significant to

- the generation of modulo number systems in which sequences are repeated, and
- an allegorical application of these modulo number systems to the orbits of bodies, whether atomic or planetary, about one another,

in which the value of a Fibonacci value F(n) squared, i.e., F(n)2 may be considered to equivalent to a value "0" in a limited number set between "1" and "F(n)²".

Universal algebraic progression sequences

To begin with, let's start with a universal sequence subject to values of a variable "n". Namely a function U(n).

I shall first introduce the concept of a Unitary Sequence [similar if not the same as an "instance of a Lucas sequence"] in which :

27

An Aether expressing Fibonacci and Lucas Sequences joined at the hip: a thought experiment

- the initial states U(1) and U(-1) of the unitary sequence U(n) are to be defined,

to be followed by

- the relationship between the initial states of the unitary sequence, defining the boundary between advancing and reversing numbers of a sequence, being defined as the "origin state".

In this case the "origin state" is defined as the difference between the two "initial states". The advancing numbers are then determined by **EQN 1** and the reversing numbers by **EQN 2**.

For the "first instance" of U(n), let´s define:

- the first "initial state" U(1) as being "1"; and
- the second "initial state" U(-1) as being "1".

Then the value of the "origin state" U(0) is defined as:

- U(1) − U(-1)
  = 1 -1
  = 0.

In this respect therefor the "first instance" of U(n), when EQN 1 and EQN 2 are applied thereto, gives rise to the "instance" of a Lucas sequence which we recognise as the Fibonacci number sequence F(n).

An Aether expressing Fibonacci and Lucas Sequences joined at the hip: a thought experiment

For a "second instance" of U(n), let´s define:

- U(1) as being "1"; and
- U(-1) as being "-1".

Then the "origin state" value of U(0) is defined as:

- U(1) – U(-1)
  = 1 – (- 1)
  = 2.

In this respect therefor the "second instance" of U(n), when EQN(1) and EQN(2) are applied thereto, gives rise to the "instance" of a Lucas sequence which we recognise as the Lucas number sequence L(n).

An Aether expressing Fibonacci and Lucas Sequences joined at the hip: a thought experiment

Basic interaction between Fibonacci and Lucas sequences to produce even numbered Fibonacci elements.

Now from the title of this book at least, the question You may be asking is,

- other than the universal application of the equations EQN 1 and EQN 2 to the respective initial conditions, in what other aspect(s) are Fibonacci and Lucas number sequences joined at the hip?

Well, the interaction between the respective values of the Fibonacci and Lucas sequences doesn't stop here. Let's take for example the Lucas sequence starting from $L(-1)$:

- $L(-1) = -1$

- $L(0) = 2\ [= 1 + 1] = L(1) - L(-1)$

- $L(1) = 1$

- $L(2) = 3\ [= 1 + 2] = L(1) + L(0)$

- $L(3) = 4\ [= 3 + 1] = L(2) + L(1)$ etc.

The values of the numbers generated by the Lucas sequence can be replicated by Fibonacci numbers. Namely:

- $L(-1) = -1 = F(-2) + F(0) = -1 + 0$

- $L(0) = 2 = F(-1) + F(1) = 1 + 1$

- $L(1) = 1 = F(0) + F(2) = 0 + 1$

An Aether expressing Fibonacci and Lucas Sequences joined at the hip: a thought experiment

- $L(2) = 3 = F(1) + F(3) = 1 + 2$

- $L(3) = 4 = F(2) + F(4) = 1 + 3$ etc.

wherein the expression for generating values of the Lucas sequence could just as easily be expressed as

- $L(n) = F(n-1) + F(n+1)$.

In this respect at least the numbers of the Fibonacci sequence appear to serve as baseline for the generation of other progressive algebraic sequences.

Now there is also a further functional relationship between numbers of the Fibonacci sequence and the Lucas sequence. Namely, if one multiplies corresponding elements $F(n)$ and $L(n)$ of the sequences with one another, one achieves a value of $F(2n)$. That is

- $F(n) \times L(n) = F(2n)$

Essentially this means that every even value element of the Fibonacci sequence (- i.e., $F(x)$) is also a product of $F(x/2)$ times $L(x/2)$.

If we take the values of "n" from 0 to 6, then we get:

- $F(0) \times L(0) = 0 \times 2 = 0 = F(0)$

- $F(1) \times L(1) = 1 \times 1 = 1 = F(2)$

- $F(2) \times L(2) = 1 \times 3 = 3 = F(4)$

31

An Aether expressing Fibonacci and Lucas Sequences joined at the hip: a thought experiment

- $F(3) \times L(3) = 2 \times 4 = 8 = F(6)$

- $F(4) \times L(4) = 3 \times 7 = 21 = F(8)$

- $F(5) \times L(5) = 5 \times 11 = 55 = F(10)$

As noted previously, the expression L(n) can also be written:

- $L(n) = F(n-1) + F(n+1)$.

Also, when it comes values of the Fibonacci sequence F(n), we can also write F(n) not only as

- $F(n) = F(n-2) + F(n-1)$,

but we can also write F(n+1) as:

- $F(n+1) = F(n) + F(n-1)$.

From here we can also then rewrite F(n) as:

- $F(n) = F(n+1) - F(n-1)$.

So, if we obtain the product of F(n) times L(n), what we obtain is:

- $F(n) \times L(n) = F(2n) = F(n+1)^2 - F(n-1)^2$.

Written another way, we can say for *even* values of "n" of the Fibonacci sequence, then:

$$L\left(\frac{n}{2}\right).F\left(\frac{n}{2}\right) = F(n) = F(\frac{n}{2} + 1)^2 - F(\frac{n}{2} - 1)^2$$

An Aether expressing Fibonacci and Lucas Sequences joined at the hip: a thought experiment

Generating the Fibonacci Sequence elements using sums and differences of Squares

Taking the whole Fibonacci sequence, it is observed that as "n" advances forward, starting from "0", the Sequence evolves:

- $F(0) = 0$ [= 1 − 1] $= F(1)^2 − F(-1)^2$

- $F(1) = 1$ [= 0 + 1] $= F(1)^2 + F(0)^2$

- $F(2) = 1$ [= 1 - 0] $= F(2)^2 − F(0)^2$

- $F(3) = 2$ [= 1 + 1] $= F(2)^2 + F(1)^2$

- $F(4) = 3$ [= 4 - 1] $= F(3)^2 − F(1)^2$

- $F(5) = 5$ [= 4 + 1] $= F(3)^2 + F(2)^2$

- $F(6) = 8$ [= 9 - 1] $= F(4)^2 − F(2)^2$

etc,

with *even* values determined by

$$F(n) = F(\frac{n}{2} + 1)^2 - F(\frac{n}{2} - 1)^2 \quad \text{[EQN 3]}$$

and with *odd* values determined by

$$F(n) = F(\frac{n+1}{2})^2 + F(\frac{n-1}{2})^2 \quad \text{[EQN 4]}$$

This also applies as "-n" reverses backward from "0". In this respect see

- $F(0) = 0$ [= 1 − 1] $= F(1)^2 − F(-1)^2$

An Aether expressing Fibonacci and Lucas Sequences joined at the hip: a thought experiment

- $F(-1) = 1 [= 0 + 1] = F(0)^2 + F(-1)^2$

- $F(-2) = -1 [= 0 - 1] = F(0)^2 - F(-2)^2$

- $F(-3) = 2 [= 1 + 1] = F(-1)^2 + F(-2)^2$

- $F(-4) = -3 [= 1 - 4] = F(-1)^2 - F(-3)^2$

- $F(-5) = 5 [= 1 + 4] = F(-2)^2 + F(-3)^2$

- $F(-6) = -8 [= 1 - 9] = F(-2)^2 - F(-4)^2$

To combine the two interlacing equations of EQN 3 and EQN 4 to provide a unitary expression for generating numbers F(n), it is proposed to introduce the use of imaginary numbers. In this way the elements of the sequence F(n) can now be expressed as interlaced sums and differences of squares:

$$F(n) = F\left(\frac{(2n + 3 + i^{2n})}{4}\right)^2 - i^{2n}F\left(\frac{(2n - 3 - i^{2n})}{4}\right)^2$$

[EQN 5]

(where "i" = square root of -1 (- i.e., $\sqrt{-1}$))

and the result thereof is, as each of the imaginary components of the product cancel each other out a Real integer number. For the sake of nomenclature, we shall call this the Aether-based Fibonacci Equation.

An Aether expressing Fibonacci and Lucas Sequences joined at the hip: a thought experiment

The generation of values of the Fibonacci Sequence as an analogy for a Tank Circuit

The basic tank circuit.

To consider the existence of an Aether which EQN(5) may possibly describe the properties thereof, let´s consider that one fundamental property of such an Aether that it should possess, namely:

> - It must be capable of transmitting Electro-Magnetic (EM) radiation (i.e., from gamma radiation to radio waves).

Aether based Tank Circuit

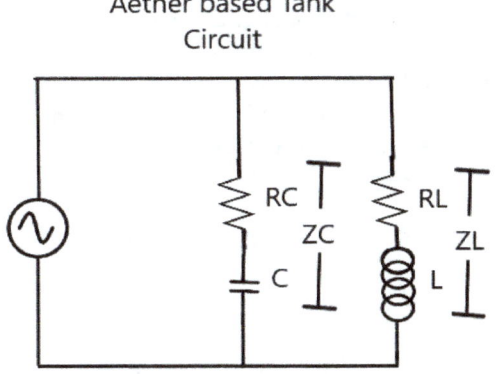

An Aether expressing Fibonacci and Lucas Sequences joined at the hip: a thought experiment

One piece of electronic technology with which electronic engineers are familiar with, that is capable of tuning EM radiation, is that of an LC filter or tank circuit [8], namely:

- a resonant circuit comprising a capacitor in parallel with an inductor of a radio transmitter / receiver.

A proposed analogy of a resistor and capacitor in parallel with a resistor and capacitor

Ordinarily for a tank circuit, only the Inductance "L" of an electronic circuit is described as comprising a resistive component. In this case however for the sake of consistency, the capacitance of this "aether-based" tank circuit is also attributed as comprising a resistive component. Thus, each of the capacitor and the inductor of such a tank circuit is expected to have a respective impedance including:

- a reactance quality, represented in electronics as -j and +j,

in series with

- a respective resistance quality.

So, for example, in this case of the Aether, assuming it can be represented by a resonant circuit analogous to that of a tank circuit:

An Aether expressing Fibonacci and Lucas Sequences joined at the hip: a thought experiment

- let the inductor circuit have an impedance of

$$ZL = RL + jL$$

- let the capacitor circuit have an impedance of

$$ZC = RC - jC.$$

This Aether resonant circuit, based on individual impedances ZL and ZC in parallel, can then have a combined impedance ZAe with

$$ZAe = \frac{ZL \times ZC}{ZL + ZC}$$

[EQN 6]

An Aether expressing Fibonacci and Lucas Sequences joined at the hip: a thought experiment

Initial conditions in the Fibonacci Sequence, factorising "0", and the number "2" as an Aether-state operator

The expansion of the Fibonacci sequence from F(0) both forwards and backwards from F(0)

As stated earlier, this interpretation of the generation of values of the Fibonacci sequence using **EQN 5** involves an expansion of the Fibonacci sequence, starting from "F(0)", in directions both forward and backward from the value when n="0" on the Real number line. It is noted that:

- i) it is only when "n" extends beyond a value greater than "2", whether on the positive or negative side of the boundary defined by F(0), that definitive values of the sequence are generated based wholly on values which have been generated before;
- ii) for values of "n" including "-2", "-1", "1" and "2", values of F(n) are generated based on squares of itself and the value F(0) at the boundary. Thus assuming no stated initial conditions exist where F(1) = 1 and F(-1) =1, then while:
  - a value of "1" is a solution for each [F(1), F(2) and F(-1), and
  - a value of "-1" is a solution for F(-2),

  it is submitted that these may not be the only solutions for these elements;

38

An Aether expressing Fibonacci and Lucas Sequences joined at the hip: a thought experiment

and

- iii) for the value of "F(0)" itself, this value is dependent on
  - the values of F(1) and F(-1) which are generated on mutually opposite sides of the boundary defined by the presence of F(0); and
  - F(0) itself insofar as the value of F(0) contributes to the values F(1) and F(-1),

  and thus, this value of "F(0)" may also have a non-defined value should neither F(1) or F(-1) be initially defined.

What can be established therefor is that it is only after values for the elements from F(-2) to F(2) are realised, that the values for the elements F(n) for "n" of three or greater can be generated in a predictable manner. The expressions for the elements of F(n) for values of "n" between -2 and 2 are reminiscent of equations to generate values of a Mandelbrot set.

An Aether expressing Fibonacci and Lucas Sequences joined at the hip: a thought experiment

A relationship between the Fibonacci sequence and the Mandelbrot set

As an illustration, let's look at the five equations of E(n) for E(-2) to E(2) for determining the values of elements F(-2) to F(2) in the case of there being no initial conditions for F(-1) and/or F(1).

i) $\quad F(-2) = F(0)^2 - F(-2)^2$
ii) $\quad F(-1) = F(0)^2 + F(-1)^2$
iii) $\quad F(0) = F(1)^2 - F(-1)^2$
iv) $\quad F(1) = F(1)^2 + F(0)^2$
v) $\quad F(2) = F(2)^2 - F(0)^2$

In the case of equation v) at least, we have an expression analogous to that of:

-   vi) $F(x) = F(x)^2 - F(y)^2$

In terms of values of "n" greater than "2" or less than "-2", the values achieved for "F(n)" are always Real and are dependent on the values "F" of previous equations also being Real. In essence there is a single solution employed to achieve each value of F(n) or F(-n). For values of "n" being equal to "-2", "-1", "1" and "2", the values of F(n) are self-referential are therefore dependent upon itself and a variable F(0). In this case also, we do not have to limit the either the values of F(2) or F(0) to being exclusively Real.

Let's adapt F(2) so that it is complex, and manage it such that it has a value:

An Aether expressing Fibonacci and Lucas Sequences joined at the hip: a thought experiment

- $F(2)_c = F(2) + iF(y)$.

In terms of Real numbers however, we maintain the equivalence on both sides of the equals sign for expression vi) so that:

- $\{F(2)_c\}Real = \{F(2)^2\}Real - \{F(0)^2\}Real$

From **EQN5** we are treating $F(0)$ per se as being the imaginary component, which is subsequently squared to give a real component $F(0)^2$.

In the case of a comparison with equation v), let's assume $F(0)$ is not equal to "0". That is, there is a disturbance in the aether whereby in determining $F(0)$, the value $F(1)^2$ is either greater than or less than $F(-1)^2$. In terms of a non-limiting example, let's assume one set of solutions for $F(2)_c$ involves $iF(y)$ of the complex value $F(2)_c$ being equal to $iF(0)$. Thus $F(2)_c$ becomes:

- $F(2)_c = F(2) + iF(0)$.

Returning to equation v), the imaginary value $i(F(0)$ must be added to both sides. Thus, expression v) becomes:

- vii) $F(2) + iF(0) = F(2)^2 - F(0)^2 + iF(0)$.

Assuming $F(2)$ is the Real component of $F(2)_c$, the expression vii) can then also be rewritten:

- $F(2) + iF(0) =$

$F(2)^2 + 2F(2).F(0) - F(0)^2 + iF(0).(1 - 2F(2))$

41

An Aether expressing Fibonacci and Lucas Sequences joined at the hip: a thought experiment

which then reduces to:

- $F(2) + iF(0) = (F(2) + iF(0))^2 + iF(0).(1 - 2F(2))$
- $F(2)_c = F(2)_c^2 + iF(0).(1 - 2F(2))$.

If we then treat the values $F(2)$ and $F(0)$ of the last portion of the above expression as being maintained at the initial values set for the values $F(2)$ and $F(0)$, namely:

- $F(2)_i$ and $F(0)_i$ respectively,

then the resulting expression vii) of:

- $F(2)_c = F(2)_c^2 + iF(0).(1 - 2F(2))$

may be considered analogous to that of the Mandelbrot equation [5]:

- $Z(x+1)_c = Z(x)_c^2 + A$

where $Z(x)_c$ is a complex variable, and A is a complex constant (in this case equal to $iF(0).(1 - 2F(2))$).

Of course, this Mandelbrot paradigm is again following the premise that $F(0)$ is not equal to zero for whatever reason. Should $F(0)$ *be equal to zero* (i.e., $F(1)^2$ and $F(-1)^2$ cancel each other), then:

- $F(2)_c = F(2)_c^2 + 0$

for which the only solution reverts to that of $F(2) = 1$ in which $F(2)$ is Real.

An Aether expressing Fibonacci and Lucas Sequences joined at the hip: a thought experiment

The Fibonacci sequence and factorising zero in an Aether-state environment.

On the other hand, let's assume initial conditions for elements F(-1) and F(1) being that of

- F(-1) and F(1) = 1.

If we attempt to determine F(0) of this Aether-based sequence when "n" = 0, then we have:

$$F(0) = F(1)^2 - (i)^{2(0)}F(-1)^2$$

[EQN 7]

If we take the above equation **EQN 7** of

$$F(0) = F(1)^2 - F(-1)^2$$

with the above initial conditions of F(1) = 1 and F(-1) = 1, then standard logic dictates that

$$F(0) = 0$$

However, it is to be noted that "n" = 0, and thus the imaginary component of F(0) becomes

- $i^0 \times i^0$

which becomes

- 1 x 1,

An Aether expressing Fibonacci and Lucas Sequences joined at the hip: a thought experiment

and thus, even when factorised, F(0) is described only in terms of Real components. Thus, if we take a literal view of the expression for F(0) of:

$$F(0) = F(1)^2 - F(-1)^2$$

then:

$$F(0) =$$

$$(F(1) + F(-1)) \times (F(1) - F(-1))$$

and

$$0 = (1 + 1) \times (1 - 1)$$

$$= (0) \times (2)$$

[EQN 8].

The "2-Aether state" as an operator" in an Aether state and its relationship to the "0-Aether state".

In terms of Aether-"states" then we achieve

- a 0-drive state =
  (2-inflection state) x
  (0-inflection state)

where:

An Aether expressing Fibonacci and Lucas Sequences joined at the hip: a thought experiment

- the "2-inflection state" forms the basis for Lucas numbers [= L(0)];

and

- the "0-inflection state" forms the basis for Fibonacci numbers [= F(0)].

This also has the consequence of a

- $2 - Inflection\ State$

$$= \frac{0\ (-\ drive\ state)}{0\ (-\ inflection\ state)}$$

or

- $2 = \dfrac{0}{0}$

As far as it is known, it is unusual to have a result for "0" divided by "0" as being any result other than "undefined". In this case however we are not considered to be dealing with "0" in terms of book-keeping in a ledger where "0" describes the absence of anything. Rather in terms of an Aether environment, it is submitted we are dealing with a "0" in terms of balancing forces at a boundary between complementary Aether expressions.

Thus, in this respect, it is proposed the value F(0) becomes either an inflection point between:

An Aether expressing Fibonacci and Lucas Sequences joined at the hip: a thought experiment

- the almost laminar behaviour of "Aethers" on the advancing side of F(0) represented by **EQN 1**, and
- the seemingly chaotic / turbulent behaviour of "Aethers" on the reversing side represented by **EQN 2**.

The "2-operator" and tidal locking about an attractor

It is further proposed that a paradigm exists of a state "2" in which, despite an object being driven in continuous motion, no change in state is perceived.

As an example of such a proposed **2-inflection state** paradigm, it is submitted that the so-called tidal-locking of the rotation of the moon to the Earth is one example thereof. That is, one side of the moon only is ever viewed from the surface of the earth despite the moon being in a state rotating about its own axis. This face of the moon being in effect static to an observer on the Earth, is nevertheless also rotating on its own axis in which different portions are illuminated by the sun.

It is proposed that a tidal-locking effect in general, based on a status "2"-inflection state, is also an effect to which electrons in atoms are also subject. That is, for electrons subject to forces from a mutual attractor to sustain a stable "orbit" about the nucleus within atoms, it is proposed that electron pairs must first operate themselves as mutually tidally locked pairs to process the Aether within the

46

An Aether expressing Fibonacci and Lucas Sequences joined at the hip: a thought experiment

boundaries of the atom as if being drawn to a common attractor. In order to remain within the confines of the atom, these tidally locked pairs of electrons are themselves subject to being drawn about another attractor which is the nucleus of the atom. In this sense a mutually tidally locked pair of electrons can be seen as being analogous to mutually tidally locked pair of planetoid bodies in the solar system, for example between the dwarf plant of Pluto and its satellite Charon.

It is also proposed that in determining behaviour of an Aether, Fibonacci number Aether states and Lucas number Aether states per se do not work in isolation, but work in tandem. They are in effect, joined at the hip.

An Aether expressing Fibonacci and Lucas Sequences joined at the hip: a thought experiment

The Cassini Identity and the Catalan Identity

Traditional expressions for the Cassini and Catalan identities

We shall now look to both the Cassini and Catalan Identities **[8]**.

The Cassini Identity

$$F(x-1).F(x+1) - F(x)^2 = (-1)^x$$

The Catalan Identity

$$F(x)^2 - F(x-r).F(x+r) = (-1)^{x-r}.F(r)^2$$

Discovery of the Cassini identity is attributed to the 17th century French astronomer Giovanni Domenico Cassini **[9]** in 1680, though the identity is presumed to have been already known to another earlier astronomer, Johannes Kepler **[10]**, in 1608.

The Catalan identity was discovered by the 19th century French mathematician Eugene Charles Catalan in 1879.

As we can see from the above, the Cassini Identity is in effect a particular case of the Catalan identity. Namely the case where *r=1*.

48

An Aether expressing Fibonacci and Lucas Sequences joined at the hip: a thought experiment

In addition, with the use of the operator $(-1)^{x-r}$, we already have a clue to the existence of the i –operator in the Fibonacci Sequence, in which $(-1)^{x-r}$ can also be expressed as $i^{2(x-r)}$.

Alternative expressions involving rewriting the Cassini and Catalan identities.

However, in further keeping with the premise of this book, it is postulated the Cassini Identity at least can also be expressed in either form of:

$$F(x).F(-x) + F(x-1).F(-(x+1)) = F(1).F(-1)$$

or

$$F(-x).F(x) + F(-(x-1)).F(x+1) = F(-1).F(1)$$

where F(-(x-1)), F(-x) and F(-(x+1)) are the respective conjugate values of F(x-1), F(x) and F(x+1).

We have thus rewritten the Cassini Identity. By extension we can also rewrite the Catalan Identity such that it includes elements of both the positive values of "X" and the corresponding negative values of "X".

An Aether expressing Fibonacci and Lucas Sequences joined at the hip: a thought experiment

However, back to the Cassini Identity, as each of F(-1) and F(1) are already defined as being equal to "1", Cassini Identity equation reduces to:

$$F(x).F(-x) - (i)^2.F(x-1).F(-(x+1)) = 1$$

or

$$F(-x).F(x) - (i)^2.F(-(x-1)).F(x+1) = 1$$

[EQN 9]

Let's try some examples with the Cassini Identity. It should be noted that the introduction of the conjugate values also means that values may be positive or negative depending upon "X" being *odd* or *even*.

For "X" = 4, then:

- F(x-1) = F(-(x-1)) = F(3) = 2
- F(x) = F(4) = 3
- F(-x) = F(-4) = -3
- F(x+1) = F(-(x+1)) = F(5) = 5

and **EQN 9** becomes:

- (3).(-3) − (-1). (2).(5) = − 9 + 10 = 1

For "X" = 7, then:

- F(x-1) = F(6) = 8

An Aether expressing Fibonacci and Lucas Sequences joined at the hip: a thought experiment

- F(-(x-1) = F(-6) = -8
- F(x) = F(-x) = F(7) = 13
- F(x+1) = F(8) = 21
- F(-(x+1)) = F(-8) = -21

and EQN 9 becomes:

- (13).(13) – (-1). (8).(-21)= 169 -168 = 1

If we now turn our attention to the Catalan Identity, it is noted that the Catalan Identity can also be expressed in different forms i.e., either one of:

$$F(x).F(-x) - (i)^{2r}.F(x-r).F(-(x+r)) = F(r).F(-r)$$

or

$$F(-x).F(x) - (i)^{2r}.F(-(x-r)).F(x+r) = F(r).F(-r)$$

[EQN 10]

Again, let's try some examples now for the Catalan Identity with variable "x" and "r".

For "X" = 5 and "r" = 2, then

- F(x-r) = F(-(x-r) = F(3) = 2
- F(x) = F(-x) = F(5) = 5
- F(x+r) = F(-(x+r)) = F(7) = 13
- F(r) = F(2) = 1
- F(-r) = F(-2) = -1

and EQN 10 becomes:

An Aether expressing Fibonacci and Lucas Sequences joined at the hip: a thought experiment

- $(5).(5) - (i)^4.(2).(13) = (1).(-1)$
- $25 - (1).(26) = -1$

For "X" = 10 and "r" = 3, then

- $F(x-r) = F(-(x-r)) = F(7) = 13$
- $F(x) = F(10) = 55$
- $F(-x) = F(-10) = -55$
- $F(x+r) = F(-(x+r)) = F(13) = 233$
- $F(r) = F(-r) = F(3) = 2$

And then EQN 10 becomes:

- $(-55).(55) - (i)^6.(13).(233) = (2).(2)$
- $-3025 - (-1).(3029) = 4$

An Aether expressing Fibonacci and Lucas Sequences joined at the hip: a thought experiment

Thoughts on a holographic universe

> [1] In the beginning God created the heaven and the earth.
>
> [2] And the earth was without form, and void; and darkness was upon the face of the deep. And the Spirit of God moved upon the face of the waters.
>
> [3] And God said, Let there be light: and there was light.
>
> [4] And God saw the light, that it was good: and God divided the light from the darkness.
>
> [5] And God called the light Day, and the darkness he called Night. And the evening and the morning were the first day.
>
> -   Genesis 1, verses 1-5

A concept of an Aether as a fabric including forward and reverse strands of time

For this exercise, let's look back at EQN 7 and interpret this considering EQN 10.

That is, with EQN 7

$$F(0) = F(1)^2 - (i)^{2(0)}F(-1)^2$$

53

An Aether expressing Fibonacci and Lucas Sequences joined at the hip: a thought experiment

let us consider EQN 10 of

$$F(x) \times F(-x) - (i)^r . F(x-r) \times (i)^r . F(-(x+r))$$

$$= F(r) \times F(-r)$$

We can rewrite F(x).F(-x) as

$$F(x).F(-x) =$$

$$F(r).F(-r) + (i)^r . F(x-r).(i)^r . F(-(x+r))$$

And the question arises as to how to interpret the behaviour of values of "r" and "X" relative to one another?

To adapt the values of "r" and "X" of rewritten EQN 10 to fit the result of EQN 7 of obtaining a value of „0", it is only required to select:

- "X" to be equal "0" and
- "r" to be equal to "1".

Then assuming:

- $F(1) = F(-1) = 1 = F(1)^2$ and
- $F(0) = F(-0) = 0 = F(0)^2$,

whereby we arrive at:

$$F(0). F(-0) =$$

$$F(1).F(-1) + (i)^1 . F(-1).(i)^1 . F(-1)$$

This implies:

54

An Aether expressing Fibonacci and Lucas Sequences joined at the hip: a thought experiment

$$F(0)^2 = F(-1).F(1) + (i)^2.F(-1)^2 \quad \text{or}$$

$$F(0) = F(1)^2 + (i)^2.F(-1)^2$$

Laser holography analogy for creating the Ever Present Now (EPN) using mutually complementary time strands.

If we look deeper at the consequences of EQN 10, then it is also seen that the value of the product $F(r).F(-r)$ is independent of any value of "X". That is, despite the oscillating values on the lefthand side of EQN 10 due to any incrementing or decrementing of values of the variable "X", the value on the righthand side stays obstinately the same at $F(r).F(-r)$.

It is proposed therefore to:

-     i) treat the variable "X" as being representative of a *"strand of time"* (for want of a better word), with "X" extending simultaneously in both a positive (future) direction and a negative (past) direction, i.e., along the Aether number line;

and to

-     ii) treat the variable "r" as a characteristic of the Aether which

55

An Aether expressing Fibonacci and Lucas Sequences joined at the hip: a thought experiment

causes the Aether number line to fold
on itself to create structure.

In this paradigm it is proposed the *"strands of time "*, in a fifth dimension, behave like beams of light split from a laser.

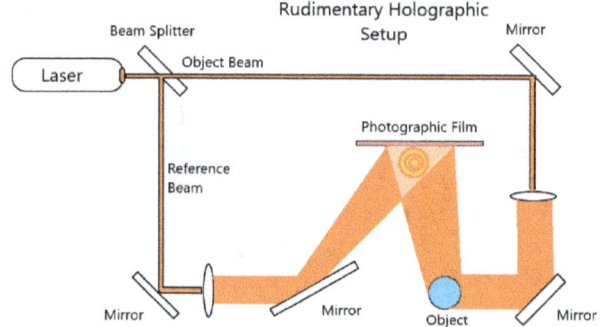

Diagram illustrating laser light being used, as an allegory for time, to generate "strands of time" including an "object time strand" and a "reference time strand" to thus obtain the Ever Present Now (EPN)

These contra-propagating *"strands of time "* then interact with each other in the Aether in a manner analogous to reference and object laser beams intersecting a holographic plate.

Thus, for the purposes of this exercise the respective Fibonacci values of F(x) and F(-x) are considered to result

An Aether expressing Fibonacci and Lucas Sequences joined at the hip: a thought experiment

from respective forward and reverse propagating *"strands of time* "of "X" and "-X" interacting with the Aether. It is then proposed that it is the instantaneous product of F(x) x F(-x) formed from these two values which provides a basis for both:

- i) physical structure in the Aether, and
- ii) the illusion of time in the Ever Present Now (EPN) to the conscious observer.

The build-up of dimensions and the mirror plane

Discussion of prerequisite conditions of F(1) and/or F(-1) defining F(0)

If we take a superficial look at the definition of F(0),

$$F(0) \quad = F(1)^2 - F(-1)^2 \quad = 1 - 1 \quad = \quad 0$$

then we must first note that either before or at "n=0", the values of F(1) and F(-1) must have already existed. That is, according to the equations:

$$F(-1) \quad = F(0)^2 + F(-1)^2 \quad = 0 + 1 \quad = \quad 1$$

$$\mathbf{F(0)} \quad = \mathbf{F(1)^2 - F(-1)^2} \quad = \mathbf{1 - 1} \quad = \quad \mathbf{0}$$

$$F(1) \quad = F(1)^2 + F(0)^2 \quad = 1 + 0 \quad = \quad 1$$

An Aether expressing Fibonacci and Lucas Sequences joined at the hip: a thought experiment

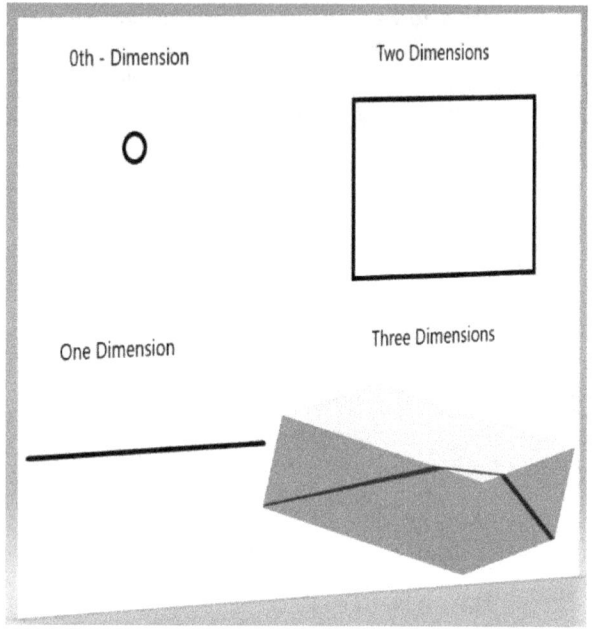

0th - Dimension

Two Dimensions

One Dimension

Three Dimensions

Illustrations of 0th to 3rd dimensions, at least as presented on a 2-dimensional plane.

Each of F(1) and F(-1) can only exist once each of the "object time strand" and "reference time strand" have already given rise to F(0). Thus, the state of F(0) here, assigned the value of "0", is really indeterminate, as are in essence values of F(1) and F(-1) in a classic chicken and egg scenario. Thus, the solution has been to assign respective values of "1" and "1" to both F(1) and F(-1). That is, insofar as F(1) and F(-1) are both dependent on the value of F(0),

An Aether expressing Fibonacci and Lucas Sequences joined at the hip: a thought experiment

and F(0) is defined as the difference between F(1) and F(-1) to obtain "0", the corresponding values U(1) and U(-1) of a universal algebraic sequence can only be indefinite values unless explicitly defined.

Relating forward and reverse time-strands to forward (object) and reverse (reference) Aether elements of Fibonacci sequences.

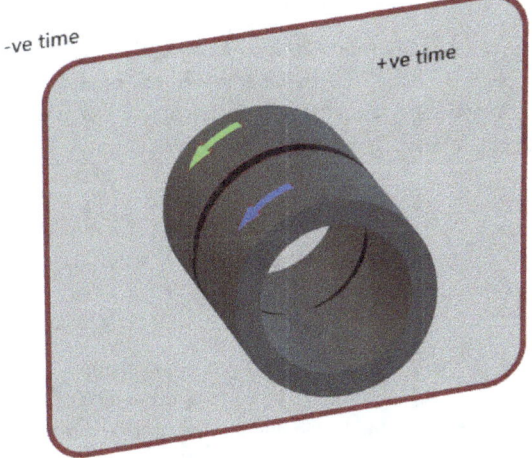

Interaction of F(-n) in green of "reference time strands" and F(+n) in blue of "object time strands" reacting with one another across the observer plane / consciousness plane to give rise to the Ever Present Now (EPN).

59

An Aether expressing Fibonacci and Lucas Sequences joined at the hip: a thought experiment

The other caveat from the above equations is that each value U(n) is obtained through the sum or difference of squares of other U(n) values. The question arises from where a complementary U(n) component may arise to interact with the original U(n) component.

Discussion of the "5th" dimension and a shadow in a holographic paradigm

It is suggested here that, in the corollary of the "photographic plate" used on the holography model, there is in essence generation of an image on both sides of the plate. F(0) at least is derived from the difference of

- the "real" number of "1" of a number line on one side of the plate moving in one direction,

interacting with

- the "complementary" / "shadow" number F(-1) on the other side of the plate moving in the other direction.

In this manner it is proposed that fifth dimensional Aethers interact with each other across the "photographic plate" to form squares of whichever polarity to both:

60

An Aether expressing Fibonacci and Lucas Sequences joined at the hip: a thought experiment

- produce bodies of Aether which express as matter in the form of atoms; and
- enable forces to be generated with which atoms in a three-dimensional space interact.

The structures and forces, which we as Beings within the Creation experience and catalogue, exhibiting properties of numbers within either a Fibonacci or Lucas sequence are then explained as epiphenomena of the fifth dimensional Aethers.

Discussion on a 0th dimension to 2nd dimension in the context of modulo number systems

As regards the 0th to 2nd dimensions, these cannot be considered supportable in a strict manner. That is, as illustrated in the graph illustrating 0th to 3rd dimensions, it is sought to make the 0th to 2nd dimensions comprehensible by drawing them as a either a point circle (0th dimension) or a mixture of lines (1st and 2nd dimensions) on a flat surface of a three-dimensional object. In this sense the 0th and 1st dimensions are not viable as they exclude the possibility of Aether interacting with "real" and "shadow" sides thereof. The 2nd dimension, while achieving Aether interaction of "real" and "shadow" sides, admit only limited interaction of Aether bodies with other Aether bodies if at all. That is any matter arising from bodies of Aether would operate only as if in one-dimensional plane. Analogous to having a modulo-1 number system in which any value equal to "1" or greater

61

An Aether expressing Fibonacci and Lucas Sequences joined at the hip: a thought experiment

would return a value of "0", these would enable possibly e.g., electrical discharge in the form of lightning strikes.

Discussion on the structure of matter in terms of an autogenerated boundary between real and shadow sides of fifth dimensional Aethers

In summary, what we as conscious Beings appear to experience is a four-dimensional world in which space is perceived as extending in all directions out to infinity. We also treat time as a fourth dimension through which bodies of matter may interact with the Aether to apply force(s) to other bodies of matter within the Aether to move within space relative to one another. I am treating matter here however as also being an expression of fifth-dimensional Aethers in which:

- a "shadow side Aether" / "reactive Aether" is bound in the form of proton-neutron bodies within the nuclei of atoms;

and in which

- a "consciousness side Aether" / "real Aether" provides for the expansion of space and allowance of communication and interaction between atoms through electromagnetism and electricity.

An Aether expressing Fibonacci and Lucas Sequences joined at the hip: a thought experiment

A short discussion on a fifth dimensional holographic Aether as part of a story of creation

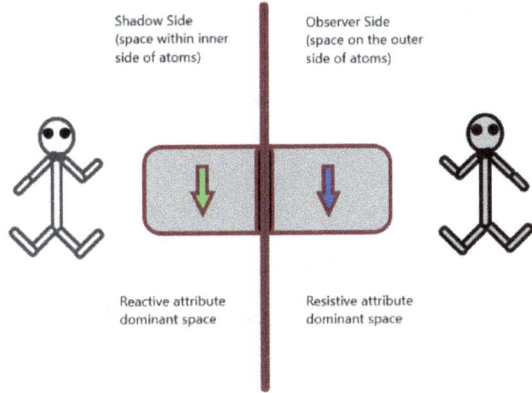

Shadow Side
(space within inner
side of atoms)

Observer Side
(space on the outer
side of atoms)

Reactive attribute
dominant space

Resistive attribute
dominant space

A representation of a separation of observer space and shadow space. Introducing a role of "consciousness" as ultimate observer, and its interaction with Aether(s) to elicit phenomena of time, matter and space through reflection in a "time-strand" induced mirror plane

I started this section with a quotation from the opening verses of Genesis. At this point it all seems rather esoteric, reminiscent of other religious texts along the lines of the bible in describing creation.

In talking of an observer side and shadow side, I am lending from the wording of the first lines of genesis:

63

An Aether expressing Fibonacci and Lucas Sequences joined at the hip: a thought experiment

1. In the beginning God created the heaven and the earth.

2. And the earth was without form, and void; and darkness was upon the face of the deep. And the Spirit of God moved upon the face of the waters.

3. And God said, Let there be light: and there was light.

4. And God saw the light, that it was good: and God divided the light from the darkness.

5. And God called the light Day, and the darkness he called Night. And the evening and the morning were the first day.

- Genesis 1:1-5

This is not to say my interpretation is the true meaning of the words written in genesis. My view of the Bible at this point is that the words are fractal. The words of genesis may prima facie refer to a process of transforming a primal state of the earth to a thriving paradise for the habitation of humanity. And the second chapter of genesis gives a second account of apparently the same creation process, or possibly the same creation process for the earth having already gone through a creation-destruction cycle. Regardless of how the creation of genesis 2 should or could be interpreted, my thinking on the creation process of genesis 1 is that it may also be considered fractal for a more fundamental process for the primal state of Aethers and the creation of time, matter, and space.

An Aether expressing Fibonacci and Lucas Sequences joined at the hip: a thought experiment

Thus, an accommodating text for this book based on the above-mentioned fifth-dimensional Aethers (and admittedly inspired from an interview given by Joseph Campbell regarding Star Wars) could go like:

- Before the beginning there was the "one", and the "one" hovered above face of the Aethers in which all was chaotic and indeterminate. Then perceiving the "one" reflected on the face of the Aethers, the "one" pronounced:
  - o let there be time / oscillation / frequency / light so that I may know that I am / I be.

The first three days of the creation story in genesis as a fractal for the creation of time, space, and matter

Thus, I propose for what it's worth the separation of observer space and shadow space as described here may correspond to the separation of light from darkness as spoken of in genesis 1.

To continue this line of thought, for the second day, genesis recounts that:

> 6 And God said, Let there be a firmament in the midst of the waters, and let it divide the waters from the waters.

65

An Aether expressing Fibonacci and Lucas Sequences joined at the hip: a thought experiment

> [7] And God made the firmament, and divided the waters which were under the firmament from the waters which were above the firmament: and it was so.
>
> [8] And God called the firmament Heaven. And the evening and the morning were the second day.
>
> o   Genesis 1:6-8

In my view or interpretation, this may be an allegory for the creation of matter and space itself, with the "firmament" corresponding to matter with the "waters" under the firmament corresponding to Aether(s) forming the internal nuclear structure of atoms, and water above the firmament corresponding to Aether(s) of what we term space. Thus, it is proposed in this paradigm that a light of creation became hidden in the self-enclosing inner Aether(s) of matter, that matter being referred to as "firmament".

An account of an atomic explosion and the light seen thereby

This is only a proposition for the relationship between matter, space, and time. But not so fanciful when one considers the accounts of witnesses to the earliest atomic

An Aether expressing Fibonacci and Lucas Sequences joined at the hip: a thought experiment

explosions in which colours beyond description are referred to. As one example [13]:

> "Ralph Carlisle Smith, the future assistant director of Los Alamos Scientific Laboratory, also observed the explosion with the naked eye. Here's what he saw:
>
> > 'I was staring straight ahead with my open left eye covered by a welders glass and my right eye remaining open and uncovered. Suddenly, my right eye was blinded by a light which appeared instantaneously all about without any build up of intensity. My left eye could see the ball of fire start up like a tremendous bubble or nob-like mushroom. I Dropped the glass from my left eye almost immediately and watched the light climb upward. The light intensity fell rapidly hence did not blind my left eye but it was still amazingly bright. It turned yellow, then red, and then beautiful purple. At first it had a translucent character but shortly turned to a tinted or colored white smoke appearance. [...]'"

[Sourced from "Surely You're Joking, Mr. Feynman!": Adventures of a Curious Character".

(Available on Amazon Kindle)]

An Aether expressing Fibonacci and Lucas Sequences joined at the hip: a thought experiment

In my opinion the above account relates to the effects of a fission process in which i.e., a God ordained separation of

- outer and inner waters (/ Aethers) from one another

and/or

- darkness from light,

as recounted in the first book of genesis, is momentarily interrupted. In my way of interpreting the tale, a firmament which defines a boundary between the nucleus of an atom and it´s electron shells is partially dispelled, just before the inner and outer Aether(s) self-seal the breach.

An Aether expressing Fibonacci and Lucas Sequences joined at the hip: a thought experiment

Modulo Number Systems (MNS) and Spin in the context of Cassini values

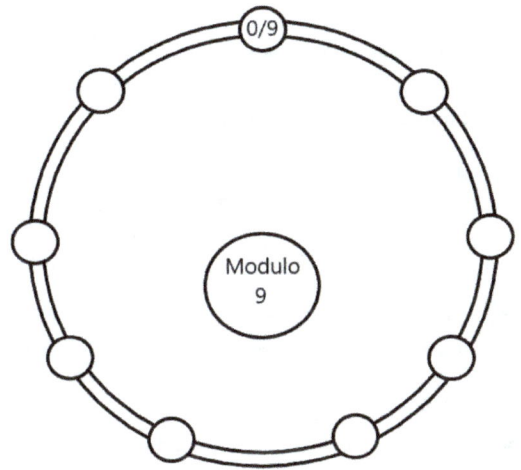

Modulo number system of F(4)² [=9]

Interpreting numbers in a modulo number system

For this exercise we will use a number system for which, if You are aware of the work of Marko Rodin, you may be familiar with. Namely the modulo-9 number system (MNS-9) in which there are nine integer states only extending from 1 through 9.

An Aether expressing Fibonacci and Lucas Sequences joined at the hip: a thought experiment

In this MNS-9 system, much like an athlete on a track running past Aether-state markers thereon, there is a simple rule:

> - once a value / distance achieved reaches or goes beyond 9 (or below -9), a value of 9 (or -9) is removed from the value / distance

and the value / distance obtained is that shown by the Aether-state marker. In addition, there is also the question of "spin". That is, the direction in which the athlete is running also determines the Aether-state of the marker.

In this corollary, there are also two possibilities or directions in which the athlete may run. Conversely there are two Aether-states by which markers may be marked according to the direction of spin. It will be further explained how these two complementary marker systems form a unitary convention of describing an Aether-state.

An Aether expressing Fibonacci and Lucas Sequences joined at the hip: a thought experiment

Positive spin convention

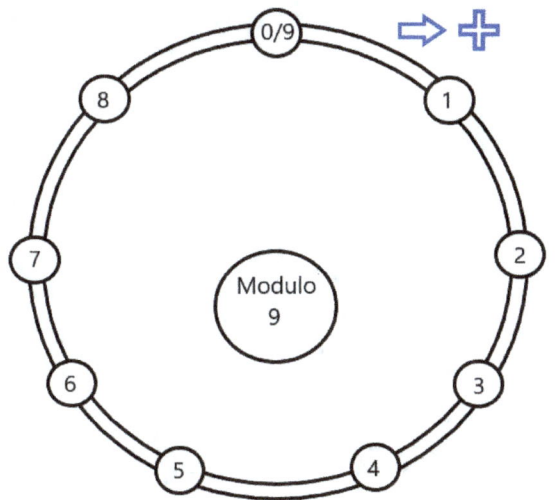

One Spin-Value of an Aether-State marker is based on the assigning of MNS-9 element values following a clockwise direction. Thus clockwise (+ve), the MNS-9 has Spin-Values of "0" to "8", after which a value of "9" reverts to "0".

An Aether expressing Fibonacci and Lucas Sequences joined at the hip: a thought experiment

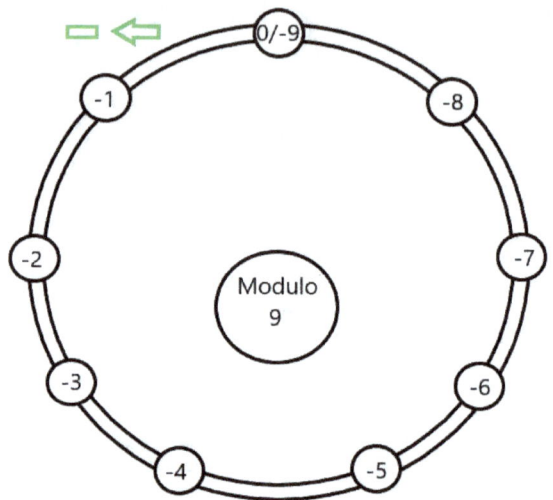

The other Spin-Value of an Aether-State is based on the assigning of MNS-9 element values following an anti-clockwise direction.

Thus, anti-clockwise (-ve), the MNS-9 has Spin-Values of "0" to "-8", after which a value of "-9" reverts to "0".

An Aether expressing Fibonacci and Lucas Sequences joined at the hip: a thought experiment

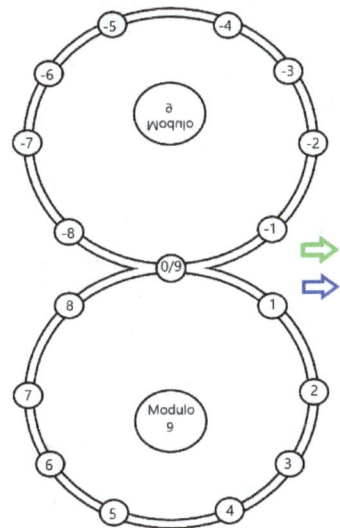

Looking at the MNS-9 circuits of mutually opposite spin directions of forward and reverse lets treat them as being copies of the same circuit, but on opposite sides of the planar surface of the equivalent "photographic plate".

Going back to the athlete on a track analogy, let's assume that the planar surface is transparent. These forward and reverse track circuits are then overlaid upon one another. Thus when viewed from different sides of the planar surface, the overlaid MNS-9 circuits provide the illusion of the athlete moving in opposite directions as s/he follows the track.

An Aether expressing Fibonacci and Lucas Sequences joined at the hip: a thought experiment

Spin correction in a Modulo Number System (MNS)

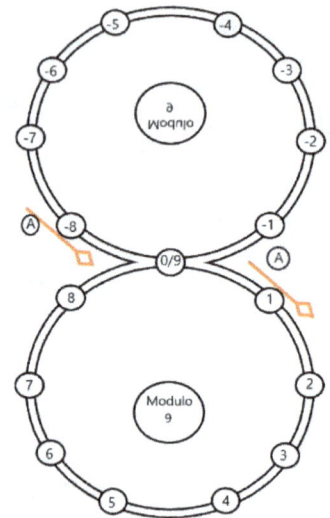

Let's now tell the athlete that s/he must run on a path following both sides of the planar surface, passing through the inflection point of "0" twice. Thus, one lap consists of two circuits of 360° giving a total of consists of 720°. If this number is familiar, it is because it is the number of degrees of rotation a fermion particle such as an electron, with a spin-½, must complete before it returns to it original state [14].

How then can we compress information to describe progress of the runner on these two circuits?

An Aether expressing Fibonacci and Lucas Sequences joined at the hip: a thought experiment

Take the athlete at station "1" of the forward circuit. The characteristic of a vector of their motion at that station is the same as that of the vector at station "-8" of the reverse circuit. Let's then call this common characteristic at stations "1" and "-8" be the vector "A".

Interpreting Aether-state values as vectors

So, for the purposes of this athlete, let:

-    Aether state -8 = Aether state 1 = vector A.

So, we repeat this exercise of assigning common vectors to each of the stations of the forward and reverse circuits such that:

-    state -7 = state 2 = vector B

-    state -6 = state 3 = vector C

-    state -5 = state 4 = vector D

-    state -4 = state 5 = vector E

-    state -3 = state 6 = vector F

-    state -2 = state 7 = vector G and

-    state -1 = state 8 = vector H.

So now, to describe the progress of the runner on the two circuits, we can describe their progress, in terms of vectors,

An Aether expressing Fibonacci and Lucas Sequences joined at the hip: a thought experiment

as going from before repeating the cycle at the inflection point of "0". However, if we now bring in the idea of symmetry of direction about the vertical plane, the information required to describe the progress of the runner may be reduced even further.

- 0-A-B-C-D-E-F-G-H to

- 0-H-G-F-E-D-C-B-A

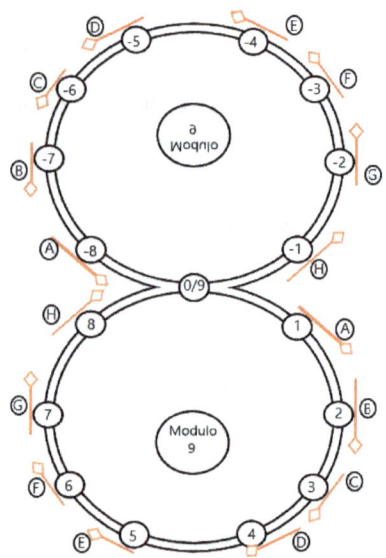

With the use of symmetry, it is further assumed that vectors:

An Aether expressing Fibonacci and Lucas Sequences joined at the hip: a thought experiment

- A and H, B and G, C and F, as well as

- D and E

are reflections of each other in terms of angle rotated relative to the vertical axis. That is, if vector A is at an angle of e.g., +60° relative to the vertical axis, then vector H is at -60°. Thus, in this context:

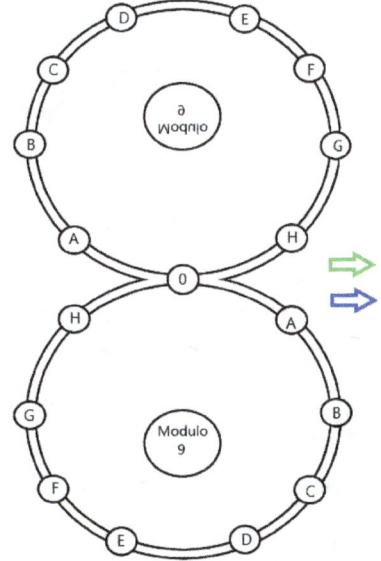

- vectors E, F, G and H can now be represented as

- vectors -D, -C, -B and –A.

An Aether expressing Fibonacci and Lucas Sequences joined at the hip: a thought experiment

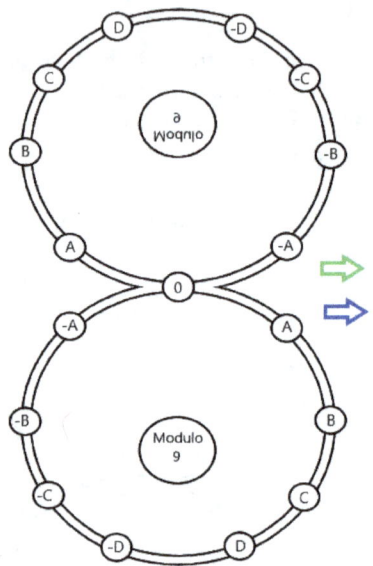

Thus, a lap of 720° having 17 stations in this MNS-9 cycle can now be described through:

- a single inflection point; and

- four variable points i.e. with their symmetrical reflections.

An Aether expressing Fibonacci and Lucas Sequences joined at the hip: a thought experiment

In spin correction, criteria for expressing an Aether-state value as either a positive value or its complementary negative value.

In this context, according to "spin correction" as exercised:

- if a value of an Aether-State in an MNS is positive, but falls in the lefthand-side of the MNS-track, the Aether-State is represented by its negative counterpart; and

- if a value of an Aether-State is negative but falls in the righthand-side of the MNS-track, the Aether-State is represented by its positive counterpart.

An Aether expressing Fibonacci and Lucas Sequences joined at the hip: a thought experiment

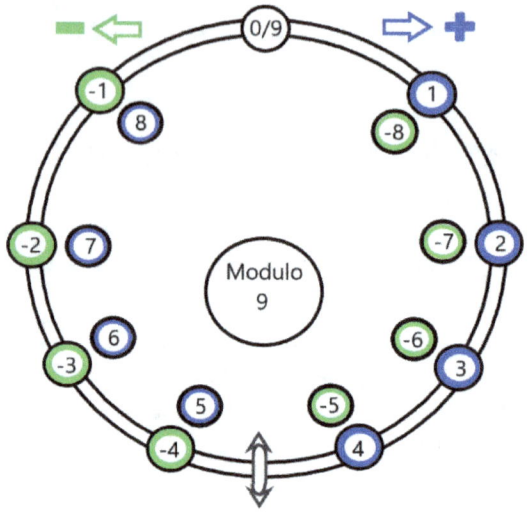

In extremes, it is submitted that it may also be said that each "Aether-state" can, in any Modulo Number System (MNS), be represented by two values, that of:

-    the forward value, and that of
-    the forward value minus the MNS value.

An Aether expressing Fibonacci and Lucas Sequences joined at the hip: a thought experiment

The expression of Fibonacci and Lucas sequences according to a Modulo Number System

Now how to put the paradigm of "Spin Correction" into practice when obtaining values of either the Fibonacci sequence, or the Lucas sequence, according to MNS-9, that is MNS-F(4)^2.

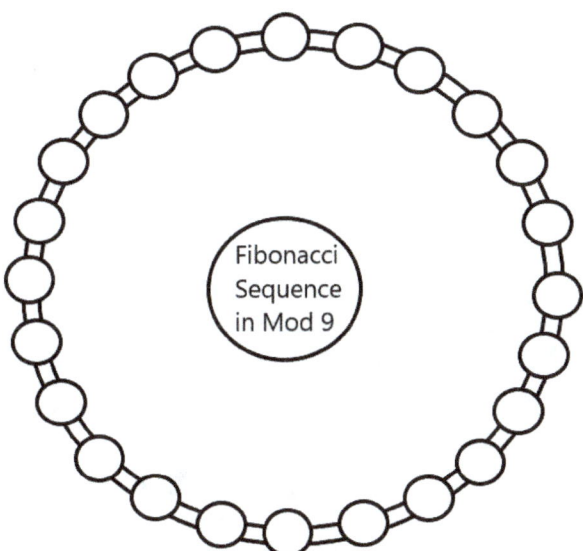

Under modulo-9, the MNS sequence has 24 values before the sequence begins repeating again.

An Aether expressing Fibonacci and Lucas Sequences joined at the hip: a thought experiment

Thus, to begin with it must be noted that:

- each of the Fibonacci Sequence and the Lucas Sequence in MNS-9 require 24 iterations before the sequence repeats; and

- different sequences of values are obtained depending on whether positive direction "n" values are sequenced or negative direction "n" values are sequenced.

An Aether expressing Fibonacci and Lucas Sequences joined at the hip: a thought experiment

Forward generated Fibonacci sequence in MNS-9

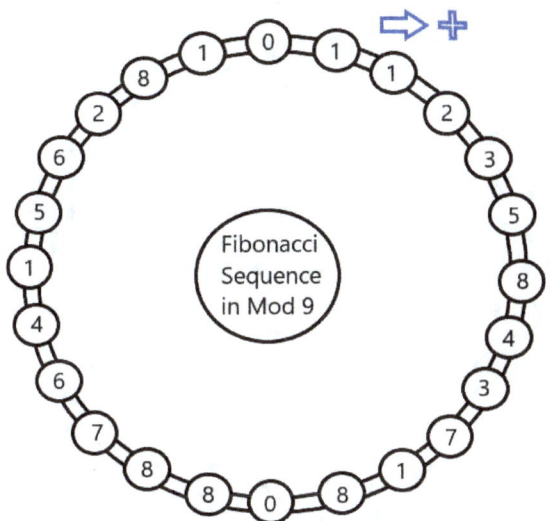

Fibonacci Sequence following the positive direction of EQN 1 for values of "n".

As one can see here below, we are generating values by adding in a direction right of "0" i.e., corresponding to the value of "F(0)" and using EQN 1.

$$\_ 5 \_ \text{-}3 \_ 2 \_ \text{-}1 \_ 1 \_ \text{"0"} \_ 1 \_ 1 \_ 2 \_ 3 \_ 5 \_$$

It follows that only positive values are produced, and that the cycle returns to the sequence "1" _ "0" _ "1" to repeat the cycle again after 24 iterations.

An Aether expressing Fibonacci and Lucas Sequences joined at the hip: a thought experiment

Reverse generated Fibonacci sequence in MNS-9

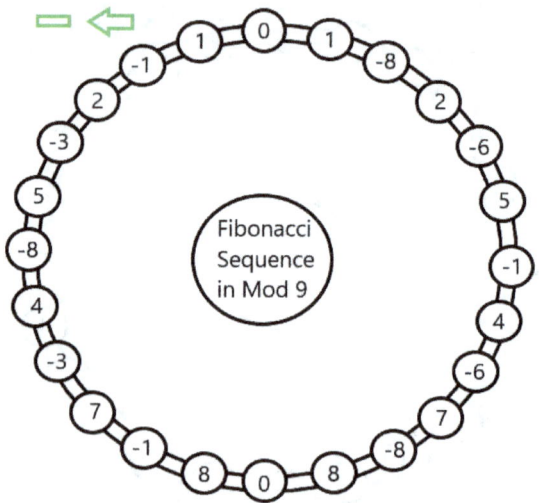

Fibonacci Sequence following the negative direction of EQN 2 for values of "-n".

As one can see here below, we are now generating values by adding in a direction left of "0" i.e., corresponding to the value of "F(0)" and using EQN 2.

_ **5** _ **-3** _ **2** _ **-1** _ **1** _ "**0**" _ 1 _ 1 _ 2 _ 3 _ 5 _

It follows that the 24 positive and negative values are produced, but that the cycle still returns to the sequence "1" _ "0" _ "1" to repeat the cycle again.

84

An Aether expressing Fibonacci and Lucas Sequences joined at the hip: a thought experiment

Spin corrected Fibonacci sequence in MNS-9

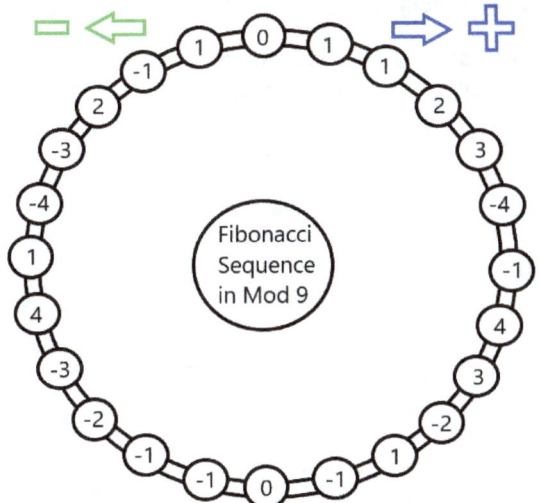

Spin corrected Fibonacci Sequence following either positive or negative direction for values of "n".

Now having been processed according to "spin correction", regardless of whether the sequence follows, from "0":

-   values of "n" in a positive direction according to EQN 1; or

-   values of "n" in a negative direction according to EQN 2,

An Aether expressing Fibonacci and Lucas Sequences joined at the hip: a thought experiment

the result of the generated Fibonacci sequence is the same sequence of numbers.

Let's now apply EQN 1 and EQN2 to the Lucas sequence where L(0) = 2 and the values of L(-1) and L(1) are "-1" and "1" respectively.

An Aether expressing Fibonacci and Lucas Sequences joined at the hip: a thought experiment

Forward and reverse generated Lucas sequences in MNS-9

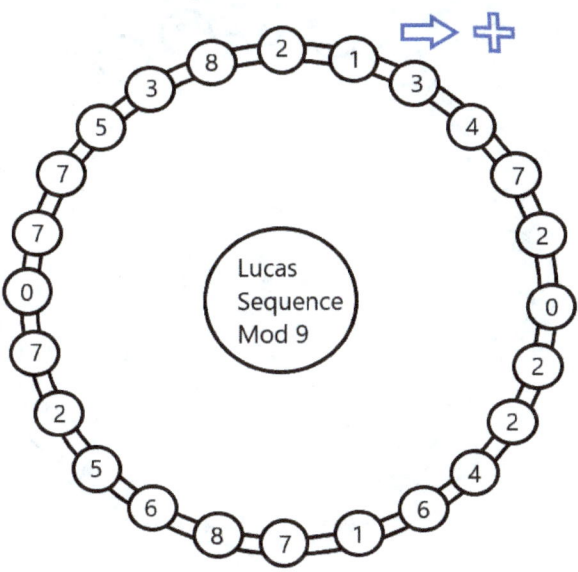

Lucas Sequence following the positive direction of EQN 1 for values of "n".

-   -11 _ 7 _ -4 _ 3 _ -1 _ "2" _ 1 _ 3 _ 4 _ 7 _ 11

An Aether expressing Fibonacci and Lucas Sequences joined at the hip: a thought experiment

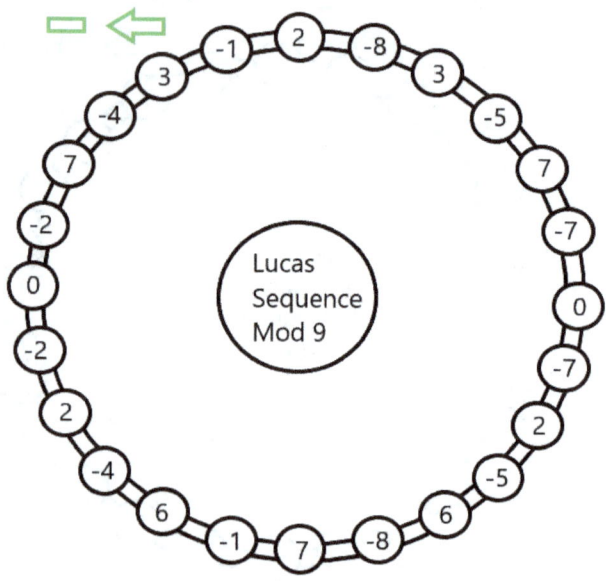

Lucas Sequence following the negative direction of EQN 2 for values of "n".

_ -11 _ 7 _ -4 _ 3 _ -1 _ "2" _ 1 _ 3 _ 4 _ 7 _ 11

As can be seen in these examples of the Fibonacci and Lucas Sequences, any number whither positive or negative when greater than or equal to "9", has "9" either added thereto or subtracted therefrom such that values (number

An Aether expressing Fibonacci and Lucas Sequences joined at the hip: a thought experiment

of Aether-states) never exceed "8". In this respect, the states "9" and "0" are synonymous.

If we now turn to the "spin corrected" Lucas sequence, we get the following.

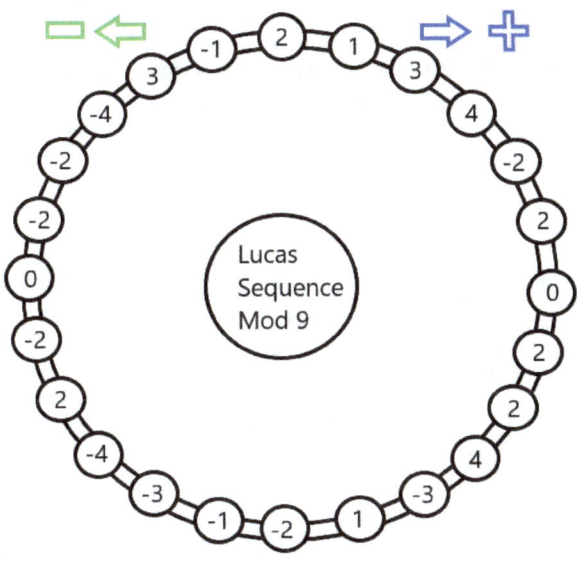

Spin corrected Lucas Sequence following either positive or negative direction for values of "n"

An Aether expressing Fibonacci and Lucas Sequences joined at the hip: a thought experiment

Now having been processed according to "spin correction", regardless of whether the Lucas sequence follows, from "2":

- values of "n" in a positive direction according to EQN 1; or

- values of "-n" in a negative direction according to EQN 2,

the result of the generated Lucas sequence is likewise the same sequence of twenty-four numbers.

An Aether expressing Fibonacci and Lucas Sequences joined at the hip: a thought experiment

A relationship between the MNS-9 spin-corrected Fibonacci sequence and the MNS-9 spin-corrected Lucas Sequence:

Comparison of the ellipse of planetary motion and the circle of a Rodin-type force diagram

As mentioned earlier, before the astronomer Cassini was accredited to have discovered the Cassini Identity in 1680, it is suspected that Johannes Kepler was already familiar with the Identity in 1608.

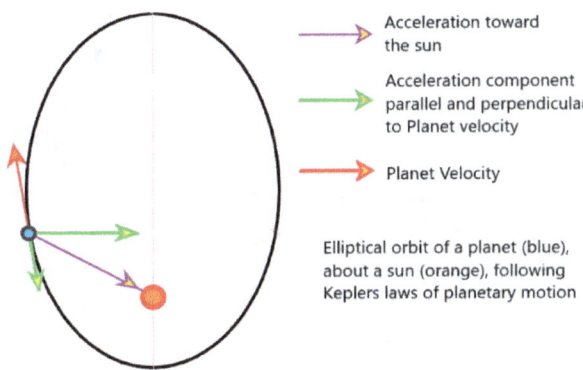

Acceleration toward the sun

Acceleration component parallel and perpendicular to Planet velocity

Planet Velocity

Elliptical orbit of a planet (blue), about a sun (orange), following Keplers laws of planetary motion

Diagram providing a rudimentary explanation of Kepler´s laws of planetary motion.

Kepler nevertheless had already become famous for many years earlier for his work on the laws of planetary motion [15]. It is unknown to this author at least as to how Kepler discovered the Cassini Identity, or how he employed the Cassini Identity.

An Aether expressing Fibonacci and Lucas Sequences joined at the hip: a thought experiment

The Cassini Identity and Marko Rodin diagrams for electrical coils

In the relatively recent past a person by the name of Marko Rodin **[16, 17]** has created somewhat of a following based on his theories regarding the efficacy numbers in the building of electrical coils.

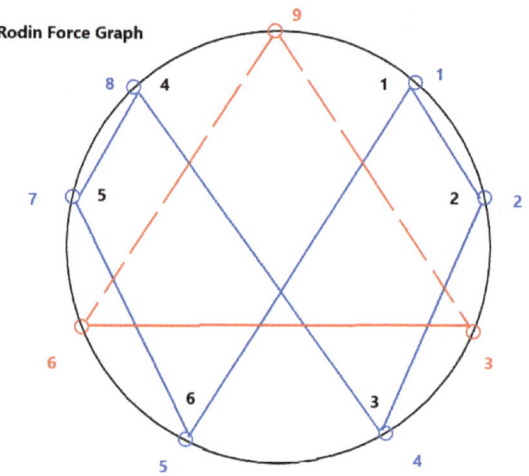

Rodin-type force diagram with multipliers "2", "5"

Especially Mr. Marko Rodin´s theories involving the use of enneagrams in a modulo-9 number system in so-called doubling and halving circuits. In a small thought experiment, let´s try to imagine if Kepler was trying to adapt a diagram along the lines of Mr. Rodin`s force

An Aether expressing Fibonacci and Lucas Sequences joined at the hip: a thought experiment

diagram to affect an ellipse. Would he have achieved something like:

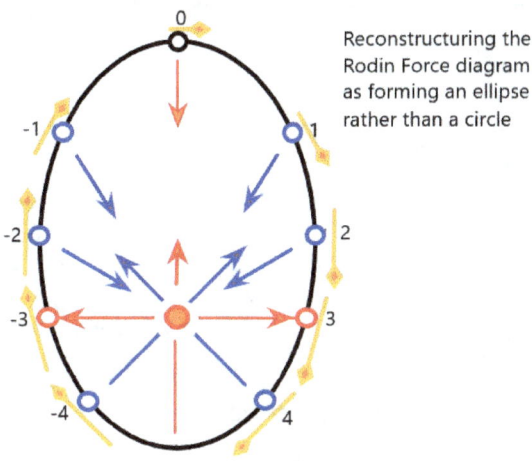

Reconstructuring the Rodin Force diagram as forming an ellipse rather than a circle

Rodin-type circle reconstructed as an ellipse.

In the Rodin force diagram across (in this case a doubling enneagram), the numbers in black merely denote the order of the sequence, the numbers in blue the component values of the MNS-9 sequence, and the numbers in red the non-participating numbers of the MNS-9. This doubling enneagram using "2" as a multiplier follows, starting from "1", a sequence of values of 1-2-4-8-7-5 [that is 2 x 8= 16 [mod-9 means 16-9 = 7]; 2 x 7 = 14 [mod-9 means 14-9 = 5]; 2 x 5 = 10 [mod-9 means 10-9 = 1]]. A converse halving

93

An Aether expressing Fibonacci and Lucas Sequences joined at the hip: a thought experiment

enneagram uses "5" as multiplier and follows, starting from "1", a sequence of values of 1-5-7-8-4-2 [that is 5 x 1 = 5; 5 x 5 =25 [mod-9 means 25-18 =7]; 5 x 7 = 35 [mod-9 means 35-27 = 8]; 5 x 8 = 40 [mod-9 means 40-36 = 4]; 5 x 4 = 20 [mod-9 means 20-18 = 2]; 5 x 2 = 10 [mod-9 means 10-9 = 1]].

The Rodin-type force diagram and the Cassini Identity giving rise to an inversion plane.

As should be noted, the value "9" of the MNS-9 sequence is $F(4)^2$ of the Fibonacci sequence, "2" is F(3) and "5" is F(5). We thus have all components of the Cassini Identity. What should also be noted is that when on multiplies the corresponding components of the doubling and halving enneagrams,

1-2-4-8-7-5 (doubling) and

1-5-7-8-4-2-1 (halving)

the resultant of each product is always "1". In effect, Mr. Rodin in demonstrating halving and doubling enneagrams is illustrating two halves of the same coin.

An Aether expressing Fibonacci and Lucas Sequences joined at the hip: a thought experiment

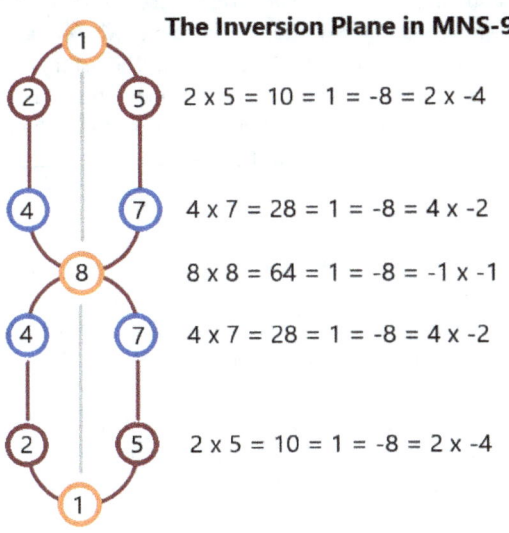

The Inversion Plane in MNS-9

$2 \times 5 = 10 = 1 = -8 = 2 \times -4$

$4 \times 7 = 28 = 1 = -8 = 4 \times -2$

$8 \times 8 = 64 = 1 = -8 = -1 \times -1$

$4 \times 7 = 28 = 1 = -8 = 4 \times -2$

$2 \times 5 = 10 = 1 = -8 = 2 \times -4$

The Cassini Identity and an inversion plane in the MNS-9 [$=F(4)^2$] circuit.

That is, in employing multipliers in an Aether-based system which follows the rules of an instance of a Lucas sequence (- in this instance the Fibonacci sequence), there appears to be a Cassini principle for MNS-$F(n)^2$ closed loop number sequences. That is each multiplier of $F(n+1)$ gives rise to a corresponding divisor of $F(n-1)$ [and *vice-versa*] such that there is unity. That is, analogous to Newtons Third law of action and reaction, for every event affecting multiplication, there is a corresponding event affecting division.

An Aether expressing Fibonacci and Lucas Sequences joined at the hip: a thought experiment

In addition, it is further submitted that an Aether-state in one MNS system may interact / resonate with either a Multiplier or a divisor derived from a MNS system having a different Modulo value if only one value of the Multiplier / Divisor of the external MNS system correlates with one of the two values ascribed to the MNS system in question.

The Catalan Identity and the inversion plane in MNS-25 [= F(5)^2] circuit

The emerging of an inversion plane does not always occur in the example given for the F(n+1) and F(n-1) Cassini Identity multipliers with respect to the MNS-9 circuit. To go back to the traditional Cassini Identity expression, to achieve an inversion plane about the value "1", the result of the difference between

- $F(n+1)*F(n-1) - F(n)^2$

must always be equal to "1". Otherwise, the resultant of each of the products will oscillate between "1" and "-1". Conversely for the Catalan identity, to achieve an inversion plane about the value "1", the result of the difference between

- $F(n+2)*F(n-2) - F(n)^2$

must always be equal to "1".

An Aether expressing Fibonacci and Lucas Sequences joined at the hip: a thought experiment

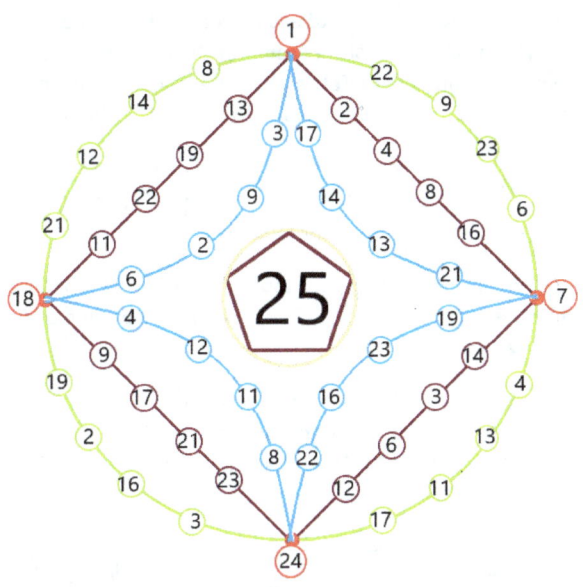

Inversion plane for modulo 25 with complementary Cassini multipliers of 8 * 22 and 3 * 17 and Catalan multipliers 2 * 13.

In this respect see the red line of the inversion plane for the MNS-F(5)^2 circuit with Catalan Identity multipliers of "2" [= F(3)] and "13" [= F(7)]. One can also adapt the Cassini Identifiers of "3" and "8" (F(4) and F(6)) to provide complementary inversion multipliers of

- F(4) * (F(5)^2 – F(6)) = "3" * "17" (blue line); and
- F(6) * (F(5)^2 – F(4)) = "8" * "22" (yellow line).

97

An Aether expressing Fibonacci and Lucas Sequences joined at the hip: a thought experiment

In addition, for elements of each colour line, one must also note in the inversion plane graph for the MNS-25 circuit that if one multiplies left to right corresponding mirror elements of a line, one obtains a value of "1". That is:

- 8 * 22 or 14 * 9 of the yellow line = "1";
- 13 * 2 or 19 * 4 of the red line = "1"; and
- 3 * 17 or 9 * 14 of the blue line = "1".

Conversely, if one multiplies top to bottom corresponding mirror elements of a colour line, one obtains a value of "-1". That is:

- 8 * 3 or 22 * 17 of the yellow line = "-1";
- 13 * 23 or 2 * 12 of the red line = "-1"; and
- 3 * 8 or 17 * 22 of the blue line = "-1".

Comparing the Fibonacci and Lucas Sequences in e.g., MNS-9 Aether circuit systems

Let's now see how, in a spin-corrected representation of the Fibonacci and Lucas sequence cycles in an MNS-9 is affected by multipliers of "2" and "5".

An Aether expressing Fibonacci and Lucas Sequences joined at the hip: a thought experiment

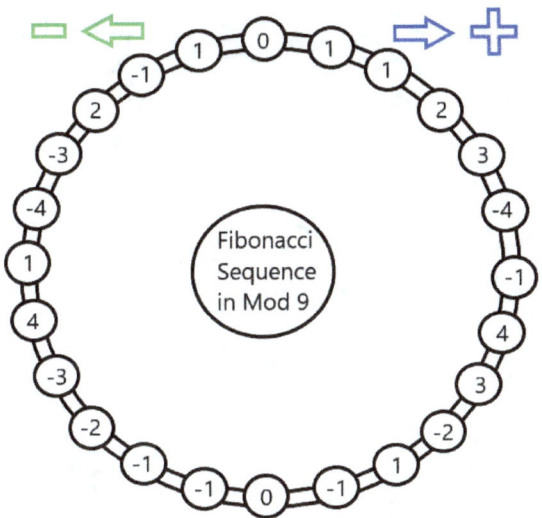

Spin corrected Fibonacci sequence in MNS-9

Here again we have the spin-corrected view of the Fibonacci sequence, with the use of EQN 1 when processing in a clockwise direction, and the use of EQN 2 in the ant-clockwise direction.

An Aether expressing Fibonacci and Lucas Sequences joined at the hip: a thought experiment

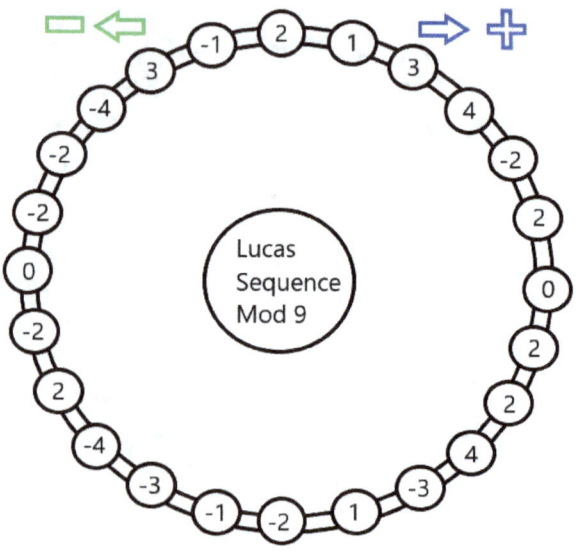

Spin corrected Lucas sequence in MNS-9

Here also we again have the spin-corrected view of the Lucas sequence, with the use of EQN 1 when processing in a clockwise direction, and the use of EQN 2 in the ant-clockwise direction.

An Aether expressing Fibonacci and Lucas Sequences
joined at the hip: a thought experiment

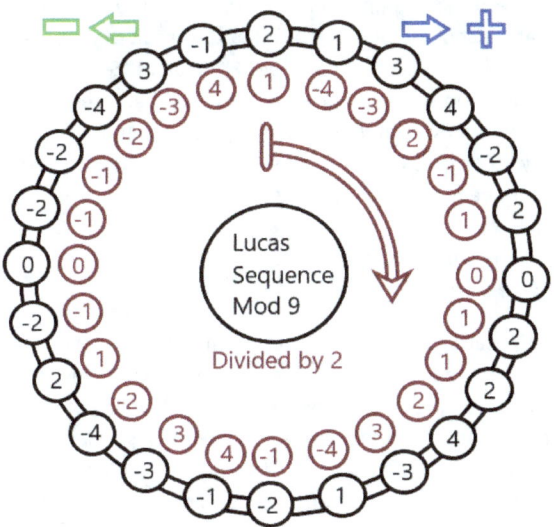

**Spin corrected Lucas sequence (outer ring) and spin
corrected Lucas sequence (inner ring) with values of Lucas
sequence divided by 2.**

Here now we again still have the spin-corrected view of the
Lucas sequence on outer side of the above sequence circle.
However, on the inner side we have the self-same values of
the Lucas sequence, but with each value divided by 2
[division by 2 is achieved in the MNS-9 circuit by either
using "2" as a divisor or "5" as a multiplier].

An Aether expressing Fibonacci and Lucas Sequences joined at the hip: a thought experiment

The "2-operator" implementing a right-hand rule.

What we should notice here is that the result of dividing values of the Lucas sequence by "2", in an MNS-9 closed number loop sequence at least, is the Fibonacci sequence but rotated by 6 values of the complete circuit of 24 values, or by 90°, to the right. Conversely, it can also be deduced that if one multiplies values of the Fibonacci sequence by a value "2", or divide by "5", one ends up with the Lucas sequence but rotated to the left by 90°.

What I hope to have demonstrated here is that the multiplication by an Aether-based value of "2" in the case of an Aether-based MNS-9 circuit at least expressing the Fibonacci sequence, achieves not only a doubling of the scalar measure of that Fibonacci value but also a left-hand rule in terms of the change of direction of any Aether-based force derived therefrom. Furthermore, what I also hope to have demonstrated is that where relationships in nature are found in terms of Lucas numbers, these may be derived from a Cassini-based multiplier interacting with a related MNS-F$(n)^2$ Fibonacci circuit sequence.

Addendum concerning the relationship between Kepler's laws of planetary motion Rodin-type graphs.

It should be noted at his point that it is only the Cassini derived multipliers / divisors of MNS-F$(4)^2$ which give rise to a diagram analogous to diagrams representing Kepler's

102

An Aether expressing Fibonacci and Lucas Sequences joined at the hip: a thought experiment

laws of planetary motion, and this by virtue of a multiplier of "2" rather than the Cassini identity per se.

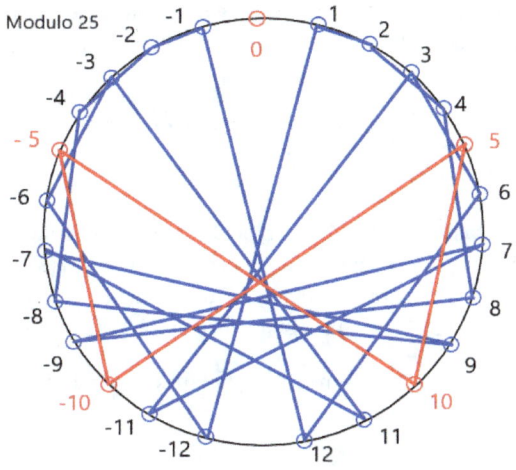

Rodin-type diagram for Modulo-25 [F(5)²] with multipliers of "2" [F(3)] and "13" [F(7)] as per Catalan Identity where "r" = 2

The Cassini and Catalan identities as applied to MNS-F(5)² circuits.

In the case of an MNS-F(5)² Aether circuit sequence involving the Cassini identity, there is more than one pair of multipliers which meet the Cassini criteria. Namely, these are of

- F(4) with F(-6) [values of "3" with "-8"] or

An Aether expressing Fibonacci and Lucas Sequences joined at the hip: a thought experiment

- F(-4) with F(6) [values of "-3" with "8"]

In multiplication sequences using non-spin corrected values this amounts to the use of two multiplier pairs:

- "3" with "25" – "8" = "3" with "17"; and
- "8" with "25" – "3" = "8" with "22".

To obtain the above Rodin-type diagram for modulo-$F(5)^2$ using the single pair of multipliers "2" and "13" [F(3) with F(7)], one requires knowledge of the Catalan identity where the value of "r" = 2. In this instance of the Catalan identity

- $F(n-2) \times F(-(n+2)) - (i)^4 \times F(n) \times F(-n) = 1$; or
- $F(3) \times F(-7) - F(5) \times F(-5) = 1$.

An Aether expressing Fibonacci and Lucas Sequences joined at the hip: a thought experiment

Multiplier M1	M1 Spin Corrected	n	Multiplier M2	M2 Spin Corrected
1	1	1	13	-12
2	2	2	19	-6
4	4	3	22	-3
8	8	4	11	11
16	-9	5	18	-7
7	7	6	9	9
14	-11	7	17	-8
3	3	8	21	-4
6	6	9	23	-2
12	12	10	24	-1
24	-1	11	12	12
23	-2	12	6	6
21	-4	13	3	3
17	-8	14	14	-11
9	9	15	7	7
18	-7	16	16	-9
11	11	17	8	8
22	-3	18	4	4
19	-6	19	2	2
13	-12	20	1	1

Table of a Rodin-type expansion for MNS-25
$[F(5)^2]$ circuit where complementary multipliers
/ divisors are "2" [F(3)] and "13" [F(-7)]

An Aether expressing Fibonacci and Lucas Sequences joined at the hip: a thought experiment

There does not appear to be however any historical evidence, as far as can be discerned, that Johannes Kepler may have also been aware of the Catalan identity as well as the Cassini identity.

An Aether expressing Fibonacci and Lucas Sequences joined at the hip: a thought experiment

Further relationships involving the Cassini Identity and the Catalan Identity

As a reminder, the Catalan Identity reads:

$$F(x).F(-x) - (i)^{2r}.F(x-r).F(-(x+r)) = F(r).F(-r)$$

or

$$F(-x).F(x) - (i)^{2r}.F(-(x-r)).F(x+r) = F(r).F(-r)$$

[EQN 10]

If we take one *component* of the above two alternatives, namely e.g.

$$- (i)^{2r}.F(x-r).F(-(x+r))$$

we can draw a table expressing the different results for values of "x" and "r", as below.

An Aether expressing Fibonacci and Lucas Sequences joined at the hip: a thought experiment

Generating tables for analysis of the relationship between variables "x" and "r" of the Catalan Identity of a balanced sequence.

n	F(n)	F(n) * (-1) * F(-n)	F(n-1) * F(-(n+1))	F(n-2) * (-1) * F(-(n+2))	F(n-3) * F(-(n+3))	F(n-4) * (-1) * F(-(n+4))	F(n-5) * F(-(n+5))
-8	-21	441					
-7	13	-169	-168				
-6	-8	64	65	63			
-5	5	-25	-24	-26	-21		
-4	-3	9	10	8	13	0	
-3	2	-4	-3	-5	0	-13	21
-2	-1	1	2	0	5	-8	26
-1	1	-1	0	-2	3	-10	24
0	0	0	1	-1	4	-9	25
1	1	-1	0	-2	3	-10	24
2	1	1	2	0	5	-8	26
3	2	-4	-3	-5	0	-13	21
4	3	9	10	8	13	0	
5	5	-25	-24	-26	-21		
6	8	64	65	63			
7	13	-169	-168				
8	21	441					

Values of Component Table for values of

$$-(i)^{2r} \times F(n-r) \times F(-(n+r))\text{ for}$$

"n" extending from "-8" to "+8" and

"r" extending from "0" to "5"

For this table:

- In the vertex substitute "n" for "x" and
- "r" extends along the hypotenuse from "0" to "5".

The values shown in the columns indicating "even" values of "r" of the table above ["r" = 0, 2, 4] are modified by multiplying by "-1". This is to ensure that the component

An Aether expressing Fibonacci and Lucas Sequences joined at the hip: a thought experiment

maintains its correct polarity characteristic in view of the product F(n-r).F(-(n+r)) also having to be modified by $(i)^{2r}$, where "i" is the square root of "-1".

Normalising values of a table with respect to values of a column

Let's now see what happens one tries to normalise one column of the component table with respect to the remaining columns.

For "r" = 0

n	F(n)	F(n) * (-1) * F(-n)	F(n-1) * F(-(n+1))	F(n-2) * (-1) * F(-(n+2))	F(n-3) * F(-(n+3))	F(n-4) * (-1) * F(-(n+4))	F(n-5) * F(-(n+5))
-3	2	0	1	-1	4	-9	25
-2	-1	0	1	-1	4	-9	25
-1	1	0	1	-1	4	-9	25
0	0	0	1	-1	4	-9	25
1	1	0	1	-1	4	-9	25
2	1	0	1	-1	4	-9	25
3	2	0	1	-1	4	-9	25

As can be see, when the values of the column identified as "F(n) * (-1) * F(-n)" are subtracted from the values of each of the columns, in each row one obtains the values of the Catalan Identity for each value of "r".

If we now repeat the normalisation of columns of the table for each value of "r", we obtain the following:

109

An Aether expressing Fibonacci and Lucas Sequences joined at the hip: a thought experiment

For "r" = 1

n	F(n)	F(n) * (-1) * F(-n)	F(n-1) * F(-(n+1))	F(n-2) * (-1) * F(-(n+2))	F(n-3) * F(-(n+3))	F(n-4) * (-1) * F(-(n+4))	F(n-5) * F(-(n+5))
-3	2	-1	0	-2	3	-10	24
-2	-1	-1	0	-2	3	-10	24
-1	1	-1	0	-2	3	-10	24
0	0	-1	0	-2	3	-10	24
1	1	-1	0	-2	3	-10	24
2	1	-1	0	-2	3	-10	24
3	2	-1	0	-2	3	-10	24

For "r" = 2

n	F(n)	F(n) * (-1) * F(-n)	F(n-1) * F(-(n+1))	F(n-2) * (-1) * F(-(n+2))	F(n-3) * F(-(n+3))	F(n-4) * (-1) * F(-(n+4))	F(n-5) * F(-(n+5))
-3	2	1	2	0	5	-8	26
-2	-1	1	2	0	5	-8	26
-1	1	1	2	0	5	-8	26
0	0	1	2	0	5	-8	26
1	1	1	2	0	5	-8	26
2	1	1	2	0	5	-8	26
3	2	1	2	0	5	-8	26

For "r" = 3

n	F(n)	F(n) * (-1) * F(-n)	F(n-1) * F(-(n+1))	F(n-2) * (-1) * F(-(n+2))	F(n-3) * F(-(n+3))	F(n-4) * (-1) * F(-(n+4))	F(n-5) * F(-(n+5))
-3	2	-4	-3	-5	0	-13	21
-2	-1	-4	-3	-5	0	-13	21
-1	1	-4	-3	-5	0	-13	21
0	0	-4	-3	-5	0	-13	21
1	1	-4	-3	-5	0	-13	21
2	1	-4	-3	-5	0	-13	21
3	2	-4	-3	-5	0	-13	21

An Aether expressing Fibonacci and Lucas Sequences joined at the hip: a thought experiment

For "r" = 4

n	F(n)	F(n) * (-1) * F(-n)	F(n-1) * F(-(n+1))	F(n-2) * (-1) * F(-(n+2))	F(n-3) * F(-(n+3))	F(n-4) * (-1) * F(-(n+4))	F(n-5) * F(-(n+5))
-3	2	9	10	8	13	0	34
-2	-1	9	10	8	13	0	34
-1	1	9	10	8	13	0	34
0	0	9	10	8	13	0	34
1	1	9	10	8	13	0	34
2	1	9	10 ·	8	13	0	34
3	2	9	10	8	13	0	34

For "r" = 5

n	F(n)	F(n) * (-1) * F(-n)	F(n-1) * F(-(n+1))	F(n-2) * (-1) * F(-(n+2))	F(n-3) * F(-(n+3))	F(n-4) * (-1) * F(-(n+4))	F(n-5) * F(-(n+5))
-3	2	-25	-24	-26	-21	-34	0
-2	-1	-25	-24	-26	-21	-34	0
-1	1	-25	-24	-26	-21	-34	0
0	0	-25	-24	-26	-21	-34	0
1	1	-25	-24	-26	-21	-34	0
2	1	-25	-24	-26	-21	-34	0
3	2	-25	-24	-26	-21	-34	0

An Aether expressing Fibonacci and Lucas Sequences joined at the hip: a thought experiment

Analysis of values generated through normalising of columns.

Let's now compare again with the original table of component values.

n	F(n)	F(n) * (-1) * F(-n)	F(n-1) * F(-(n+1))	F(n-2) * (-1) * F(-(n+2))	F(n-3) * F(-(n+3))	F(n-4) * (-1) * F(-(n+4))	F(n-5) * F(-(n+5))
-8	-21	441					
-7	13	-169	-168				
-6	-8	64	65	63			
-5	5	-25	-24	-26	-21		
-4	-3	9	10	8	13	0	
-3	2	-4	-3	-5	0	-13	21
-2	-1	1	2	0	5	-8	26
-1	1	-1	0	-2	3	-10	24
0	0	0	1	-1	4	-9	25
1	1	-1	0	-2	3	-10	24
2	1	1	2	0	5	-8	26
3	2	-4	-3	-5	0	-13	21
4	3	9	10	8	13	0	
5	5	-25	-24	-26	-21		
6	8	64	65	63			
7	13	-169	-168				
8	21	441					

Component Table of the Catalan Identity for the Fibonacci sequence

On examination, one can see that the values of the columns of each normalising table correspond / correlate to the values of the rows of the normalising column in question.

An Aether expressing Fibonacci and Lucas Sequences joined at the hip: a thought experiment

It appears therefor, for the Fibonacci sequence at least, that the Catalan is a case or instance of a higher-level equation. That of:

$$(i)^{2a}.F(x-a).F(-(x+a)) - (i)^{2r}.F(x-r).F(-(x+r))$$

$$= (i)^{2a}.F(r-a).F(-(r+a))$$

or

$$(i)^{2a}.F(-(x-a)).F(x+a) - (i)^{2r}.F(-(x-r)).F(x+r)$$

$$= (i)^{2a}.F(-(r-a)).F(r+a)$$

[EQN 11]

where "a" is any integer value greater than "0". And so, the Catalan Identity is the case of this equation for the condition "a" = 0.

The case of the Lucas sequence and normalising columns of tables for the Catalan Identity

Having provided an example of a component table for the Fibonacci Sequence, we shall now see an example of such a table for the Lucas sequence.

An Aether expressing Fibonacci and Lucas Sequences joined at the hip: a thought experiment

n	F(n)	F(n) * (-1) * F(-n)	F(n-1) * F(-(n+1))	F(n-2) * (-1) * F(-(n+2))	F(n-3) * F(-(n+3))	F(n-4) * (-1) * F(-(n+4))	F(n-5) * F(-(n+5))
-8	47	-2209					
-7	-29	841	846				
-6	18	-324	-319	-329			
-5	-11	121	126	116	141		
-4	7	-49	-44	-54	-29	-94	
-3	-4	16	21	11	36	-29	141
-2	3	-9	-4	-14	11	-54	116
-1	-1	1	6	-4	21	-44	126
0	2	-4	1	-9	16	-49	121
1	1	1	6	-4	21	-44	126
2	3	-9	-4	-14	11	-54	116
3	4	16	21	11	36	-29	141
4	7	-49	-44	-54	-29	-94	
5	11	121	126	116	141		
6	18	-324	-319	-329			
7	29	841	846				
8	47	-2209					

Component Table of the Catalan Identity for the Lucas sequence

For this component table we may also carry out normalising procedures for each column of the table.

For "r" = 0

n	F(n)	F(n) * (-1) * F(-n)	F(n-1) * F(-(n+1))	F(n-2) * (-1) * F(-(n+2))	F(n-3) * F(-(n+3))	F(n-4) * (-1) * F(-(n+4))	F(n-5) * F(-(n+5))
-3	-4	0	5	-5	20	-45	125
-2	3	0	5	-5	20	-45	125
-1	-1	0	5	-5	20	-45	125
0	2	0	5	-5	20	-45	125
1	1	0	5	-5	20	-45	125
2	3	0	5	-5	20	-45	125
3	4	0	5	-5	20	-45	125

An Aether expressing Fibonacci and Lucas Sequences joined at the hip: a thought experiment

For "r" = 1

n	F(n)	F(n) * (-1) * F(-n)	F(n-1) * F(-(n+1))	F(n-2) * (-1) * F(-(n+2))	F(n-3) * F(-(n+3))	F(n-4) * (-1) * F(-(n+4))	F(n-5) * F(-(n+5))
-3	-4	-5	0	-10	15	-50	120
-2	3	-5	0	-10	15	-50	120
-1	-1	-5	0	-10	15	-50	120
0	2	-5	0	-10	15	-50	120
1	1	-5	0	-10	15	-50	120
2	3	-5	0	-10	15	-50	120
3	4	-5	0	-10	15	-50	120

For "r" = 2

n	F(n)	F(n) * (-1) * F(-n)	F(n-1) * F(-(n+1))	F(n-2) * (-1) * F(-(n+2))	F(n-3) * F(-(n+3))	F(n-4) * (-1) * F(-(n+4))	F(n-5) * F(-(n+5))
-3	-4	5	10	0	25	-40	130
-2	3	5	10	0	25	-40	130
-1	-1	5	10	0	25	-40	130
0	2	5	10	0	25	-40	130
1	1	5	10	0	25	-40	130
2	3	5	10	0	25	-40	130
3	4	5	10	0	25	-40	130

For "r" = 3

n	F(n)	F(n) * (-1) * F(-n)	F(n-1) * F(-(n+1))	F(n-2) * (-1) * F(-(n+2))	F(n-3) * F(-(n+3))	F(n-4) * (-1) * F(-(n+4))	F(n-5) * F(-(n+5))
-3	-4	-20	-15	-25	0	-65	105
-2	3	-20	-15	-25	0	-65	105
-1	-1	-20	-15	-25	0	-65	105
0	2	-20	-15	-25	0	-65	105
1	1	-20	-15	-25	0	-65	105
2	3	-20	-15	-25	0	-65	105
3	4	-20	-15	-25	0	-65	105

An Aether expressing Fibonacci and Lucas Sequences joined at the hip: a thought experiment

For "r" = 4

n	F(n)	F(n) * (-1) * F(-n)	F(n-1) * F(-(n+1))	F(n-2) * (-1) * F(-(n+2))	F(n-3) * F(-(n+3))	F(n-4) * (-1) * F(-(n+4))	F(n-5) * F(-(n+5))
-3	-4	45	50	40	65	0	170
-2	3	45	50	40	65	0	170
-1	-1	45	50	40	65	0	170
0	2	45	50	40	65	0	170
1	1	45	50	40	65	0	170
2	3	45	50	40	65	0	170
3	4	45	50	40	65	0	170

For "r" = 5

n	F(n)	F(n) * (-1) * F(-n)	F(n-1) * F(-(n+1))	F(n-2) * (-1) * F(-(n+2))	F(n-3) * F(-(n+3))	F(n-4) * (-1) * F(-(n+4))	F(n-5) * F(-(n+5))
-3	-4	-125	-120	-130	-105	-170	0
-2	3	-125	-120	-130	-105	-170	0
-1	-1	-125	-120	-130	-105	-170	0
0	2	-125	-120	-130	-105	-170	0
1	1	-125	-120	-130	-105	-170	0
2	3	-125	-120	-130	-105	-170	0
3	4	-125	-120	-130	-105	-170	0

As can be seen here, the resultant normalising of columns does not bring a corresponding representation of the rows of the component table for the column in question being displayed in the columns of the normalising tables. What one achieves is a repeat of the normalising tables of the Fibonacci sequence, but with each number multiplied by the value "5". Namely the value of the Cassini value of the Lucas sequence itself.

116

An Aether expressing Fibonacci and Lucas Sequences joined at the hip: a thought experiment

Defining properties of balanced and unbalanced sequences
This relationship between the Cassini Identity and the tables of the normalised components of the Catalan identity is only true however for algebraic sequences along the lines of the Fibonacci sequence [assuming a Cassini Identity of "1"] and the Lucas Sequence, as well as their harmonics. In short, a *"balanced"* algebraic sequence having an inflection point ["0" or "2"] in which numbers on one side of the inflection point have values corresponding to numbers on the other side of the inflection point. For example, with:

- "r" = 0
- $F(0) = 0$ and
- $F(1) = 7$ wherein the Cassini Identity = -49

the first normalised table below is obtained:

n	F(n)	F(n) * (-1) * F(-n)	F(n-1) * F(-(n+1))	F(n-2) * (-1) * F(-(n+2))	F(n-3) * F(-(n+3))	F(n-4) * (-1) * F(-(n+4))	F(n-5) * F(-(n+5))
-3	14	0	49	-49	196	-441	1225
-2	-7	0	49	-49	196	-441	1225
-1	7	0	49	-49	196	-441	1225
0	0	0	49	-49	196	-441	1225
1	7	0	49	-49	196	-441	1225
2	7	0	49	-49	196	-441	1225
3	14	0	49	-49	196	-441	1225

Likewise, with:

- "r" = 0
- $L(0) = 14$ and
- $L(1) = 7$ wherein the Cassini Identity = 245

the following first normalised table is obtained:

An Aether expressing Fibonacci and Lucas Sequences joined at the hip: a thought experiment

n	F(n)	F(n) * (-1) * F(-n)	F(n-1) * F(-(n+1))	F(n-2) * (-1) * F(-(n+2))	F(n-3) * F(-(n+3))	F(n-4) * (-1) * F(-(n+4))	F(n-5) * F(-(n+5))
-3	-28	0	245	-245	980	-2205	6125
-2	21	0	245	-245	980	-2205	6125
-1	-7	0	245	-245	980	-2205	6125
0	14	0	245	-245	980	-2205	6125
1	7	0	245	-245	980	-2205	6125
2	21	0	245	-245	980	-2205	6125
3	28	0	245	-245	980	-2205	6125

In both cases a direct relationship between the Catalan identity and the Cassini Identity, for this harmonic of the Fibonacci sequence at least, can be readily seen. Further, it is submitted, insofar as:

- Values in a column do not change with any change in the value of the variable "n", and
- Changes in the value of "n" may be analogous to the passing of time,

that these values of the "component table" may indicate that Aether(s) subject to conditions reflecting Catalan identity of **EQN 12** support the maintaining of a structure *per se*. It is submitted that this "structure" may be in terms of e.g., elements constituting the nucleus of an atom or a relationship between objects maintaining an orbit with one another.

An Aether expressing Fibonacci and Lucas Sequences joined at the hip: a thought experiment

Examples of unbalanced sequences

Now investigating a case of an "unbalanced" algebraic sequence with

- "r" = 0
- U(0) = 1 and
- U(1) = 6 wherein the Cassini Identity = -29

the following is a resulting component table:

n	F(n)	F(n) * (-1) * F(-n)	F(n-1) * F(-(n+1))	F(n-2) * (-1) * F(-(n+2))	F(n-3) * F(-(n+3))	F(n-4) * (-1) * F(-(n+4))	F(n-5) * F(-(n+5))
-8	-92	12788					
-7	57	-4902	-4865				
-6	-35	1855	1892	1840			
-5	22	-726	-689	-741	-644		
-4	-13	260	297	245	342	-139	
-3	9	-117	-80	-132	-35	-516	368
-2	-4	28	65	13	110	-371	513
-1	5	-30	7	-45	52	-429	455
0	1	-1	36	-16	81	-400	484
1	6	-30	7	-45	52	-429	455
2	7	28	65	13	110	-371	513
3	13	-117	-80	-132	-35	-516	368
4	20	260	297	245	342	-139	
5	33	-726	-689	-741	-644		
6	53	1855	1892	1840			
7	86	-4902	-4865				
8	139	12788					

and the following first normalising table is produced:

n	F(n)	F(n) * (-1) * F(-n)	F(n-1) * F(-(n+1))	F(n-2) * (-1) * F(-(n+2))	F(n-3) * F(-(n+3))	F(n-4) * (-1) * F(-(n+4))	F(n-5) * F(-(n+5))
-3	9	0	37	-15	82	-399	485
-2	-4	0	37	-15	82	-399	485
-1	5	0	37	-15	82	-399	485
0	1	0	37	-15	82	-399	485
1	6	0	37	-15	82	-399	485
2	7	0	37	-15	82	-399	485
3	13	0	37	-15	82	-399	485

An Aether expressing Fibonacci and Lucas Sequences joined at the hip: a thought experiment

In this case, if there is any straight-forward relationship with the values of the Catalan identity of the Fibonacci sequence and the Cassini Identity, this relationship cannot be perceived.

A higher expression of the Catalan identity for balanced algebraic sequences

Considering the above therefor, if I wish to write a corresponding universal equation U(n) for a higher expression of the Catalan identity for balanced algebraic sequences, I propose it would have to look something like this:

$$(i)^{2a}.U(x-a).U(-(x+a)) - (i)^{2r}.U(x-r).U(-(x+r))$$
$$= (i)^{2a}.(CI_U).F(r-a).F(-(r+a))$$

or

$$(i)^{2a}.U(-(x-a)).U(x+a) - (i)^{2r}.U(-(x-r)).U(x+r)$$
$$= (i)^{2a}.(CI_U).F(-(r-a)).F(r+a)$$

[EQN 12]

where CI_U is the absolute value of the Cassini Identity for the balanced algebraic sequence in question.

An Aether expressing Fibonacci and Lucas Sequences joined at the hip: a thought experiment

The Cassini Identity and component tables involving unbalanced algebraic sequences.

So up to now I have been discussing "component tables" based on the Catalan Identity for balanced sequences along the lines of the Fibonacci sequence and the Lucas sequence. These sequences have a structure so:

Fibonacci Sequence

-    5 _ -3 _ 2 _ -1 _ 1 _ **0** _ 1 _ 1 _ 2 _ 3 _ 5

Lucas Sequence

-11 _ 7 _ -4 _ 3 _ -1 _ **2** _ 1 _ 3 _ 4 _ 7 _ 11

In both these instances there is a defined central inflection point value in which the amplitude of values in sequence on one side is reflected in the self-same sequence of amplitudes on the other side (even if there is switching of polarities in some cases). The "component values" of these component tables in these instances are based on the "component" of the Catalan Identity for a sequence U(x) of:

$$- (i)^{2r}.U(-(x-r)).U(x+r)$$

or

$$- (i)^{2r}.U(-(x-r)).U(x+r)$$

of EQN 12.

121

An Aether expressing Fibonacci and Lucas Sequences joined at the hip: a thought experiment

Generating tables for analysis of the relationship between variables "x" and "r" of the Catalan Identity of an unbalanced sequence.

For unbalanced sequences, there is only an inferred central inflection point value in which the amplitude of values in sequence on one side is differs from the sequence of amplitudes on the other side. An example of this may be:

$$12 \_ -7 \_ 5 \_ -2 \_ 3 \_ 1 \_ 4 \_ 5 \_ 9 \_ 14 \_ 23$$

In this case the value "1" is the inferred inflection point as it has the minimum amplitude within the sequence. For creating the component table in this case however, the "component" is based on the expression:

$$F(x - r).F(x + r)$$

of a modified-lite Catalan Identity equation of

$$U(x - r).U(x + r) - U(x)^2 = U(r)^2$$

rather than

$$- (i)^{2r}.F(x - r).F(-(x + r))$$

of **EQN 10**.

Thus, for the unbalanced sequence of:

- $$12 \_ -7 \_ 5 \_ -2 \_ 3 \_ 1 \_ 4 \_ 5 \_ 9 \_ 14 \_ 23$$

the resulting component table comprises:

122

An Aether expressing Fibonacci and Lucas Sequences joined at the hip: a thought experiment

n	F(n)	F(n) * F(n)	F(n-1) * F(n+1)	F(n-2) * F(n+2)	F(n-3) * F(n+3)	F(n-4) * F(n+4)	F(n-5) * F(n+5)
-8	-50	2500					
-7	31	961	950				
-6	-19	361	372	350			
-5	12	144	133	155	100		
-4	-7	49	60	38	93	-50	
-3	5	25	14	36	-19	124	-250
-2	-2	4	15	-7	48	-95	279
-1	3	9	-2	20	-35	108	-266
0	1	1	12	-10	45	-98	276
1	4	16	5	27	-28	115	-259
2	5	25	36	14	69	-74	300
3	9	81	70	92	37	180	-194
4	14	196	207	185	240	97	
5	23	529	518	540	485		
6	37	1369	1380	1358			
7	60	3600	3589				
8	97	9409					

and the corresponding normalised tales comprise:

For "r" = 0

n	F(n)	F(n).F(n)	F(n-1) * F(n+1)	F(n-2) * F(n+2)	F(n-3) * F(n+3)	F(n-4) * F(n+4)	F(n-5) * F(n+5)
-3	5	0	-11	11	-44	99	-275
-2	-2	0	11	-11	44	-99	275
-1	3	0	-11	11	-44	99	-275
0	1	0	11	-11	44	-99	275
1	4	0	-11	11	-44	99	-275
2	5	0	11	-11	44	-99	275
3	9	0	-11	11	-44	99	-275

An Aether expressing Fibonacci and Lucas Sequences joined at the hip: a thought experiment

For "r" = 1

n	F(n)	F(n).F(n)	F(n-1) * F(n+1)	F(n-2) * F(n+2)	F(n-3) * F(n+3)	F(n-4) * F(n+4)	F(n-5) * F(n+5)
-3	5	11	0	22	-33	110	-264
-2	-2	-11	0	-22	33	-110	264
-1	3	11	0	22	-33	110	-264
0	1	-11	0	-22	33	-110	264
1	4	11	0	22	-33	110	-264
2	5	-11	0	-22	33	-110	264
3	9	11	0	22	-33	110	-264

For "r" = 2

n	F(n)	F(n).F(n)	F(n-1) * F(n+1)	F(n-2) * F(n+2)	F(n-3) * F(n+3)	F(n-4) * F(n+4)	F(n-5) * F(n+5)
-3	5	-11	-22	0	-55	88	-286
-2	-2	11	22	0	55	-88	286
-1	3	-11	-22	0	-55	88	-286
0	1	11	22	0	55	-88	286
1	4	-11	-22	0	-55	88	-286
2	5	11	22	0	55	-88	286
3	9	-11	-22	0	-55	88	-286

For "r" = 3

n	F(n)	F(n).F(n)	F(n-1) * F(n+1)	F(n-2) * F(n+2)	F(n-3) * F(n+3)	F(n-4) * F(n+4)	F(n-5) * F(n+5)
-3	5	44	33	55	0	143	-231
-2	-2	-44	-33	-55	0	-143	231
-1	3	44	33	55	0	143	-231
0	1	-44	-33	-55	0	-143	231
1	4	44	33	55	0	143	-231
2	5	-44	-33	-55	0	-143	231
3	9	44	33	55	0	143	-231

An Aether expressing Fibonacci and Lucas Sequences joined at the hip: a thought experiment

For "r" = 4

n	F(n)	F(n).F(n)	F(n-1) * F(n+1)	F(n-2) * F(n+2)	F(n-3) * F(n+3)	F(n-4) * F(n+4)	F(n-5) * F(n+5)
-3	5	-99	-110	-88	-143	0	-374
-2	-2	99	110	88	143	0	374
-1	3	-99	-110	-88	-143	0	-374
0	1	99	110	88	143	0	374
1	4	-99	-110	-88	-143	0	-374
2	5	99	110	88	143	0	374
3	9	-99	-110	-88	-143	0	-374

For "r" = 5

n	F(n)	F(n).F(n)	F(n-1) * F(n+1)	F(n-2) * F(n+2)	F(n-3) * F(n+3)	F(n-4) * F(n+4)	F(n-5) * F(n+5)
-3	5	275	264	286	231	374	0
-2	-2	-275	-264	-286	-231	-374	0
-1	3	275	264	286	231	374	0
0	1	-275	-264	-286	-231	-374	0
1	4	275	264	286	231	374	0
2	5	-275	-264	-286	-231	-374	0
3	9	275	264	286	231	374	0

Analysis of values generated through normalising of columns.

Thus, in this case again, considering what is illustrated in the normalised component tables, we can rewrite the "modified-lite Catalan Identity" equation as:

$$U(x - r).U(x + r) - U(x - a).U(x + a)$$

$$= (-1)^{x-r}.(CI_U).F(a - r).F(a + r)$$

125

An Aether expressing Fibonacci and Lucas Sequences joined at the hip: a thought experiment

Thus, the output can still be traced back to the primal Fibonacci sequence times the absolute value of the Cassini Identity. In essence, for unbalanced sequences we can also recreate in part the results of the component table for EQN 12 in which balanced sequences are used. The difference between the tables being that:

- elements of the "component table" of unbalanced sequences are based on multipliers $U(x + r) \times U(x - r)$ about a value $U(x)$

rather than

- elements of the "component table" of balanced sequences are based on e.g. multipliers $U(x + r) \times U(-(x-r))$ about an inflection point of $F(0)$ or $L(0)$.

The Cassini Identity and Rodin-type force graphs for unbalanced sequences

As illustrated earlier with squares of values of the Fibonacci sequence, one can generate based on values derived from the Cassini Identity a Rodin-type force diagram so:

126

An Aether expressing Fibonacci and Lucas Sequences joined at the hip: a thought experiment

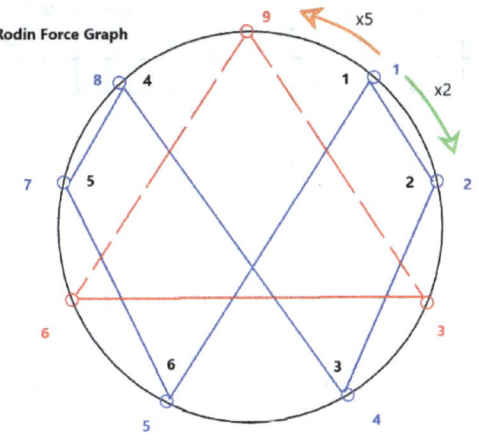

Rodin-type Force Graph for F(4)² illustrating contra-directional multipliers of F(3) [=2]] and F(5) [=5]

An example table of a sequences of Aether values in a balanced MNS-9 graph generated by Cassini Identity multipliers.

As can be seen here, the two multipliers "2" [M1] and "5" [M2] derived from the values of the Cassini Identity for "n" = 4 give rise to the following mutually complementary sequences:

An Aether expressing Fibonacci and Lucas Sequences joined at the hip: a thought experiment

Multiplier M1	M1 Spin Corrected	n	Multiplier M2	M2 Spin Corrected
1	1	1	5	-4
2	2	2	7	-2
4	4	3	8	-1
8	-1	4	4	4
7	-2	5	2	2
5	-4	6	1	1

Values obtained by multipliers M1= 2 and M2=5

In this case, thanks in part for the multiplier M1 being equal to "2", values followed by the sequence complete a circuit about the MNS-9 circuit following the sequence 1 _ 2_ 4 _ 8 _ 7 _ 5. What is to be noted however is that this sequence about the MNS-9 circuit avoids either:

- the root of "9" [namely "3"]
- any multiple of the root of "9" [namely "6"]; or
- "9" itself.

The value F(n) of an MNS-F(n)^2 number system as a dual value of root "0"

In this respect the value "3" [= F(4)] and multiples thereof are considered to have characteristics like that of dual numbers **[7]**. That is, if two numbers having these values of "3" are multiplied by one another, the value obtained being "9" or an integer multiple thereof would inherently be equal to zero. As of writing, in contrast to the behaviour of values obtained from the Rodin-Force diagram, I am not

128

An Aether expressing Fibonacci and Lucas Sequences joined at the hip: a thought experiment

aware of any oscillation behaviour [- cf. oscillation between "3" and "6" of Rodin Force graph] being ascribed to dual numbers when a single dual number is multiplied by any other element value of the MNS-9 circuit which is not a dual number.

One thing that we must also note in respect of the MNS-9 circuit at least is that I am treating, in terms of the Cassini Identity and the Catalan Identity, the value $F(4)^2$ or "9" of the MNS-$F(4)^2$ or MNS-9 circuit as being equivalent to "$F(0)$" or "0" at the heart of a "balanced" algebraic sequence. In this respect the behaviour of Aether vortices derived from squares of values of elements of the Fibonacci sequence is considered to be a fractal of an Aether vortex having $F(0)$ at the centre thereof, especially as the individual components of the squared-value $F(n)^2$ are considered to be obtained from mutually opposite sides i.e., $F(n)$ and $F(-n)$, of the inflection point $F(0)$ to give a value $F(n) * F(-n)$.

Determining Cassini Identity multipliers in an unbalanced algebraic sequence

If we now turn our attention to algebraic sequences which are unbalanced, we have a completely different relationship between:

- Multipliers of the Cassini identity about the value $U(n)$ of the sequence i.e., those multipliers of
    - $U(n+1)$ and $U(n-1)$
  used to generate the Cassini identity value

129

An Aether expressing Fibonacci and Lucas Sequences joined at the hip: a thought experiment

and the Cassini Identity value itself. Rather than the values of the MNS circuit of the Rodin-type graph being generated based on U(n)², if one generates a modulo number system based on the value of the Cassini identity itself, multiplication by the values U(n+1) and U(n-1) recreates values of the unbalanced sequence, in sequence.

Let's take for example the unbalanced sequence:

$$12 \_ -7 \_ 5 \_ -2 \_ 3 \_ 1 \_ 4 \_ 5 \_ 9 \_ 14 \_ 23$$

The Cassini Identity value obtained amounts to

- $3 \times 4 - 1 = 11$

If we take U(n) as being either e.g., -7 or 5, we would have sets of multiplier values of either:

- 12, 5 and -7, -2 respectively assuming that multipliers are obtained from each side of the assigned value U(n); or
- 12, 9 or 5, 23 and -7, 14 or -2, 5 respectively assuming that
    - a) the value "1" could be assigned the inflection value; and
    - b) the multiplier values are obtained from complementary sides of complementary values on each side of the inferred inflection value; and
    - c) the MNS circuit is based on squares of one or other of U(n) and U(-n) and not a product of U(n) x U(-n).

An Aether expressing Fibonacci and Lucas Sequences joined at the hip: a thought experiment

Prima facie however, these sets of values as multipliers do not give rise to sets of element value sequences which complement one another, even if we could assume the inflection point occurs about the value in the sequence having the lowest magnitude.

An Aether expressing Fibonacci and Lucas Sequences joined at the hip: a thought experiment

Examples of Tables and Rodin-type graphs for unbalanced algebraic sequences.

If, however a Rodin-type force graph is generated in which the Cassini Identity itself can be used as the value giving rise to vortex centre in an MNS-Cl$_U$ modulo system, the following occurs:

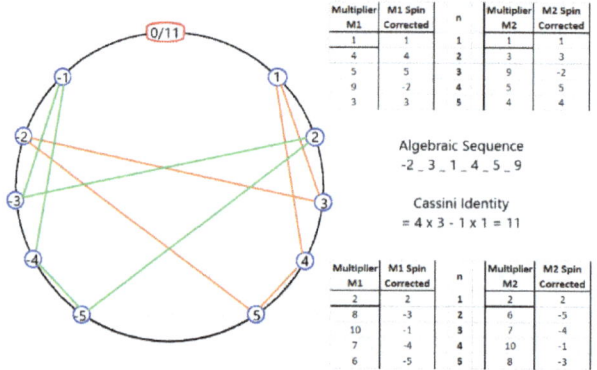

Multiplier M1	M1 Spin Corrected	n	Multiplier M2	M2 Spin Corrected
1	1	1	1	1
4	4	2	3	3
5	5	3	9	-2
9	-2	4	5	5
3	3	5	4	4

Algebraic Sequence
-2 _ 3 _ 1 _ 4 _ 5 _ 9

Cassini Identity
= 4 x 3 - 1 x 1 = 11

Multiplier M1	M1 Spin Corrected	n	Multiplier M2	M2 Spin Corrected
2	2	1	2	2
8	-3	2	6	-5
10	-1	3	7	-4
7	-4	4	10	-1
6	-5	5	8	-3

Rodin-type force graph for MNS-11 circuit using multipliers "3" and "4" about an inflection point of "1" in an algebraic sequence

-7 _ 5 _ [-2 _ 3 _ 1 _ 4 _ 5] _ 9 _ 14

An Aether expressing Fibonacci and Lucas Sequences joined at the hip: a thought experiment

The generation of source unbalanced algebraic sequence in Rodin-type graphs employing complementary Cassini Identity multipliers

As one sees from the graph and multiplier tables:

Multiplier M1	M1 Spin Corrected	n	Multiplier M2	M2 Spin Corrected
1	1	1	3	3
4	4	2	9	-2
5	5	3	5	5
9	-2	4	4	4
3	3	5	1	1

Sequences for multipliers 4 [M1] and 3 [M2] starting from "1" (orange line).

the result of the line sequence illustrated in orange in the above graph is a repeat of the original sequence [or at least that part of the sequence in brackets]. That is each of the multipliers recreate the bracketed portion of the sequence,

- one in a clockwise direction ["4" as M1],
- the other in an anti-clock direction ["3" as M2].

What is also noted in this case is that only half of the elements of the MNS-11 circuit are used in each closed sequence. One must introduce a second set of values obtained by multiplying a value "2" by 4 and 3 to obtain the second closed complementary sequence shown in green.

133

An Aether expressing Fibonacci and Lucas Sequences joined at the hip: a thought experiment

Multiplier M1	M1 Spin Corrected	n	Multiplier M2	M2 Spin Corrected
2	2	1	6	-5
8	-3	2	7	-4
10	-1	3	10	-1
7	-4	4	8	-3
6	-5	5	2	2

Sequences for multipliers 4 [M1] and 3 [M2] starting from "2"

In the above Rodin-type graph it appears the Cassini identity multipliers give rise to two complementary but independent systems operating in the same MNS-11 circuit. Alone these sequences appear unbalanced, but together they exhibit a symmetry about a centre.

Now to adjust the system where one of the multipliers "4" is divided by "2" so that the M1 multiplier is the operator "2". The other multiplier M2 is also modified by multiplying by "2" to give a value of "6". This gives the following result:

An Aether expressing Fibonacci and Lucas Sequences joined at the hip: a thought experiment

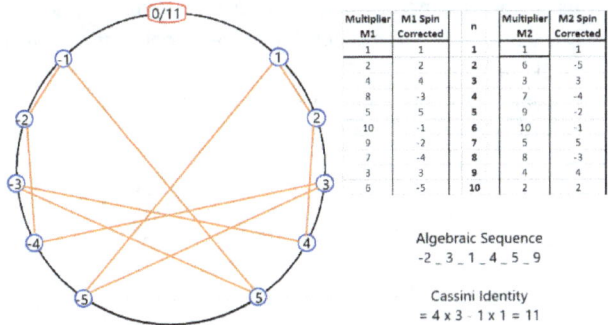

Algebraic Sequence
-2 _ 3 _ 1 _ 4 _ 5 _ 9

Cassini Identity
= 4 x 3 - 1 x 1 = 11

Rodin-type Force graph for MNS-11 with multipliers "2" [M1] and "6" [M2] producing a single sequence.

In other words, a pair of complementary multipliers / divisors in which one multiplier ("2") traverses the MNS-11 circuit in a clockwise direction and the other divisor ("6") in an anti-clockwise direction.

Multiplier M1	M1 Spin Corrected	n	Multiplier M2	M2 Spin Corrected
1	1	1	6	-5
2	2	2	3	3
4	4	3	7	-4
8	-3	4	9	-2
5	5	5	10	-1
10	-1	6	5	5
9	-2	7	8	-3
7	-4	8	4	4
3	3	9	2	2
6	-5	10	1	1

Multipliers 2 [M1] and 6 [M2] for MNS-11 circuit.

An Aether expressing Fibonacci and Lucas Sequences joined at the hip: a thought experiment

As can be seen here, when one uses the "2" operator in the MNS-11 circuit, there is now a complete sequence in which all elements of the MNS-11 circuit are included, except for the value "11" itself (in effect the acting "0" attractor at the centre of the MNS-11 circuit).

If we take it one step further for the sequence

$$-7\_ 5 \_ [-2 \_ 3 \_ 1 \_ 4 \_ 5] \_ 9 \_ 14$$

we can still treat "1" as the inferred inflection point and use multipliers of "5" and "-2" [or "9"] as equivalents to U(x + 2) and U(x – 2) as would be used in a Catalan-like expression. For this the following complementary sequences are obtained:

Multiplier M1	M1 Spin Corrected	n	Multiplier M2	M2 Spin Corrected
1	1	1	9	-2
5	5	2	4	4
3	3	3	3	3
4	4	4	5	5
9	-2	5	1	1

Multipliers 5 [M1] and 9 [M2] for MNS-11 circuit.

An Aether expressing Fibonacci and Lucas Sequences joined at the hip: a thought experiment

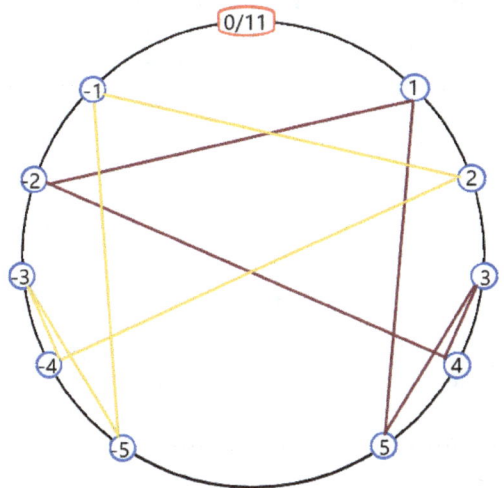

Multipliers "5" and "-2" [= 9] about an inflection point of "1" in MNS-11 circuit

We note here that the sequence of the numbers of the circuit for these multipliers "5" and "9" also form an algebraic sequence. Not $U(n+1) = U(n) + U(n-1)$ as per the Fibonacci sequence, but rather a harmonic thereof of:

-   $U(n+1) = U(n) + U(n-2)$

That is, starting with "n" = 3 with multiplier "5" in this MNS-11 circuit, we obtain the following generated circuit sequence of

-   1 _ 5 _ 3 _ 4 _ 9

An Aether expressing Fibonacci and Lucas Sequences joined at the hip: a thought experiment

which can be repeated in algebraic progression form so:

- $F(n + 1)$ = $F(n) + F(n-2)$

That is:

- 3   + 1   =   4 [3 x 5 =
  15 [- 11] = 4]
- 4   + 5   =   9 [4 x 5 =
  20 [- 11] = 9]
- 9[= -2] + 3   =   1 [9 x 5 =
  45 [- 44] = 1]
- 1   + 4   =   5 [1 x 5 =
  5]
- 5   + 9   =   3 [5 x 5 =
  25 [- 22] = 3]

As such, through the value of the Cassini Identity acting as an operator, it is submitted that:

- once a sequence is established as a "habit" about an Aether-type nexus point acting as an attractor,

then the sequence may be maintained over time with minimum computational or Aether processing effort.

An Aether expressing Fibonacci and Lucas Sequences joined at the hip: a thought experiment

Equivalence of results from multiplication and addition in algebraic sequence building in MNS-CI circuits of unbalanced sequences

So how can we express this relationship between results of multiplication and algebraic mathematically?

If an algebraic equation gives rise to a sequence

- $U(I-3)-U(I-2)-U(I-1)-U(I)-U(I+1)-U(I+2)-U(I+3)$

which generates a Cassini Identity value of

- $CI = U(I+1) * U(I-1) - U(I)^2$

where:

- $U(I)$ is the value of the inferred inflection point
- $U(I+1)$ is the value ahead of $U(I)$
- $U(I-1)$ is the value behind $U(I)$,

and I would generate a value $U(I+a+1)$ where "a" is an integer value, then in an MNS-CI circuit an equivalence arises of

- $U(I+a+1) = U(I+a) * U(I+x) = U(I+a) + U(I+a-x)$

where $U(I+x)$ is an Xth element ahead or behind the value $U(I)$.

An Aether expressing Fibonacci and Lucas Sequences joined at the hip: a thought experiment

As another example concerning the Cassini identity, the following illustrates a Rodin-type force graph for a MNS-29 sequence of

$$22 \_ -13 \_ 9 \_ -4 \_ 5 \_ 1 \_ 6 \_ 7 \_ 13 \_ 20$$

giving rise to a MNS-29 circuit with multipliers "5" and "6".

M1 Spin Corrected	M2 Spin Corrected	n	M1 Spin Corrected	M2 Spin Corrected
1	1	1	2	2
5	6	2	10	12
-4	7	3	-8	14
9	13	4	-11	-3
-13	-9	5	3	11
-7	4	6	-14	8
-6	-5	7	-12	-10
-1	-1	8	-2	-2
-5	-6	9	-10	-12
4	-7	10	8	-14
-9	-13	11	11	3
13	9	12	-3	-11
7	-4	13	14	-8
6	5	14	12	10

Algebraic Sequence
$$22 \_ -13 \_ 9 \_ -4 \_ 5 \_ 1 \_ 6 \_ 7 \_ 13 \_ 20$$
Cassini Identity
$$6 \times 5 - 1 \times 1 = 29$$

An Aether expressing Fibonacci and Lucas Sequences joined at the hip: a thought experiment

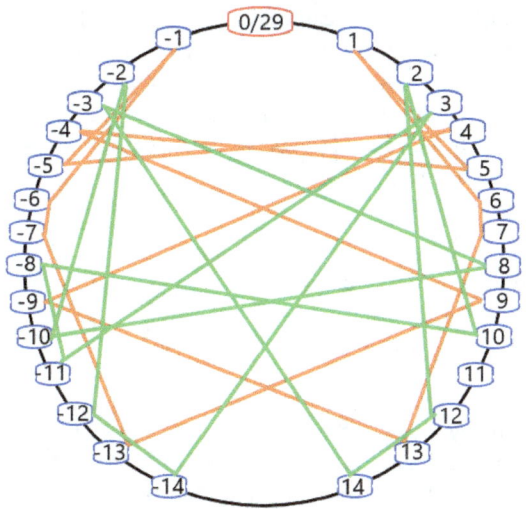

Rodin-type force graph for MNS-29 circuit using multipliers "5" and "6" about an inflection point of "1" in an algebraic sequence:

-    **22 _ -13 _ 9 _ -4 _ 5 _ 1 _ 6 _ 7 _ 13 _ 20**

Likewise for the Cassini identity value of 29, the M2 value ["6"] under modulo 29 follows the sequence in a clockwise direction while the M1 value ["5"] follows the sequence in an anti-clockwise direction. Furthermore, the sequences generated also follow an algebraic sequence of $U(n+1) = U(n) + U(n-1)$

141

An Aether expressing Fibonacci and Lucas Sequences joined at the hip: a thought experiment

In conclusion regarding unbalanced sequences, it is further noted that with the above two examples for each of the Cassini Identity values of "11" and "29", that these values themselves are Lucas sequence values, with the value "11" being L(5) and value "29" being L(7) respectively in the Lucas sequence.

Equivalence of results from multiplication and addition in algebraic sequence building in MNS-$F(n)^2$ circuits of balanced sequences

As already shown in the case of $F(4)^2$, two mutually complementary sequences are generated for multipliers F(3) and F(5). The sequences are generated in an MNS-9 circuit based on:

- $U(n+1) = U(n) * F(3)$ or
- $U(n-1) = U(n) * F(5)$

and gives the respective results, starting from "1" as:

- 1_2_4_8_7_5 and
- 1_5_7_8_4_2.

An Aether expressing Fibonacci and Lucas Sequences
joined at the hip: a thought experiment

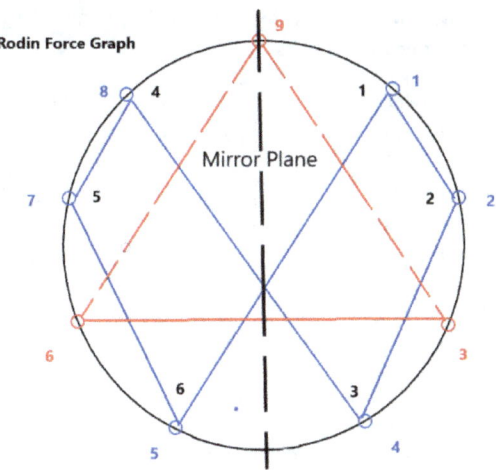

If we wish to regenerate the sequences through an
algebraic addition rather than through multiplication, we
note that if shift the first sequence with respect to itself
and add the elements, we get the following results:

- 1 [U(1)]   +   7 [U(5)]   =   8 [U(4)]
- 2 [U(2)]   +   5 [U(6)]   =   7 [U(5)]
- 4 [U(3)]   +   1 [U(1)]   =   5 [U(6)]
- 8 [U(4)]   +   2 [U(2)]   =   1 [U(1)]
- 7 [U(5)]   +   4 [U(3)]   =   2 [U(2)]
- 5 [U(6)]   +   8 [U(4)]   =   4 [U(3)]

From here we can see that this sequence is regenerated
through the expression:

- $U(n) = U(n-3) + U(n-5)$.

An Aether expressing Fibonacci and Lucas Sequences joined at the hip: a thought experiment

Alternately, if we subtract the elements from one another, we get the following results:

-	1 [U(1)]	-	2 [U(2)]	=	8 [U(4)]
-	2 [U(2)]	-	4 [U(3)]	=	7 [U(5)]
-	4 [U(3)]	-	8 [U(4)]	=	5 [U(6)]
-	8 [U(4)]	-	7 [U(5)]	=	1 [U(1)]
-	7 [U(5)]	-	5 [U(6)]	=	2 [U(2)]
-	5 [U(6)]	-	1 [U(1)]	=	4 [U(3)]

we can see that this sequence is also regenerated through the expression:

- $U(n) = U(n-3) - U(n-2)$

As such for the forward cycling sequence at least, we find that in this MNS-F(4)^2 circuit, the sequence generated by

- $U(n) = U(n-1) * F(3)$

Can be recreated by either one of

- $U(n) = U(n - F(4)) + U(n - F(5))$

or

- $U(n) = U(n - F(4)) - U(n - F(3))$.

Prima facie the above two expressions for generating the forward cycling sequence are reminiscent of the expression for the Cassini Identity. That is, if one multiplies the sequences, one obtains:

- $U(n)^2 = U(n-3)^2 + U(n-3) * [U(n-5) - U(n-2)]$

144

An Aether expressing Fibonacci and Lucas Sequences joined at the hip: a thought experiment

$$- U(n-2) * U(n-5)$$

which for each combination of values from the MNS-9 circuit, gives a value of "0" for all values of "n".

If we take another example of a balanced sequence , in the case of $F(5)^2$, two mutually complementary sequences are also generated for multipliers F(3) and F(7). The sequences are generated in an MNS-25 circuit based on expressions:

- $U(n+1) = U(n) * F(3)$ or
- $U(n-1) = U(n) * F(7)$

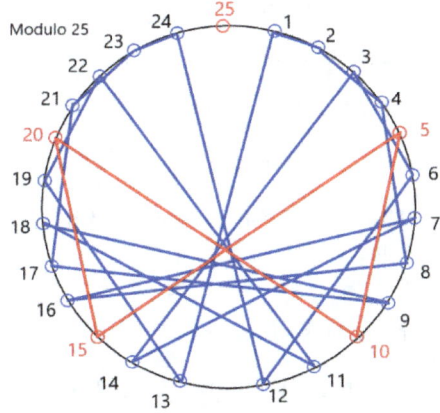

Graph based on multipliers F(7) and F(3) in Catalan Identity F(7)*F(3) - F(5)^2 = 1

An Aether expressing Fibonacci and Lucas Sequences joined at the hip: a thought experiment

Here we have the forward cycling sequence for F(3) being a multiplier of:

- 1 _ 2 _ 4 _ 8 _ 16 _ 7 _ 14 _ 3 _ 6 _ 12
  24_23_21_17 _ 9 _ 18 _ 11 _ 22 _ 19 _ 13

Extending the teaching from the example with MNS-9, it is expected that U(n) can be predicted by the expression:

- U(n) = U(n - F(5)) + U(n - F(7)) and
- U(n) = U(n - F(5)) - U(n - F(3)).

However, as it turns out, the expressions required are:

- U(n) = U(n - F(5)) + U(n + F(3)) and
- U(n) = U(n - F(5)) - U(n + F(7)).

So, for this example, it turns out that for

- U(1) = 1,

we can generate from values:

- 18 [U(16 [= 21-5])] + 8 [U(4 [= 1+3])]
  = 26
  = 1 [mod 25]

And from the converse example

- 18 [U(16 [= 21-5])] - 17 [U(14 [= 1+13])]
  = 1

If we progress further, then:

- U(2) = U(17) + U(5) or - U(15)

146

An Aether expressing Fibonacci and Lucas Sequences joined at the hip: a thought experiment

$$= 11 + 16 \text{ or } - 9$$

$$= 2$$

As we can see here, even with the balanced sequence of the Fibonacci sequence, to generate a Rodin-type force graph, multiplication is not required. Rather, the circuit values generated through constant multiplication by a component of the Cassini Identity or Catalan Identity [with the caveat that the result of the Cassini Identity or Catalan Identity = "1" as of these examples], can be recreated through addition or attraction of elements of the MNS-$F(n)^2$ circuit as part of an algebraic progression. Furthermore, the elements of the MNS-$F(n)^2$ circuit used in the algebraic progression are determined based on the multipliers used in the Cassini or Catalan Identities themselves. In summary, for *even* values of "n" for MNS-$F(n)^2$ circuits, sequences can be recreated through algebraic progression by either one of

- $U(n) = U(n - F(n)) + U(n - F(n+1))$

and

- $U(n) = U(n - F(n)) - U(n - F(n-1))$.

And for *odd* values of "n" for MNS-$F(n)^2$ circuits, sequences can be recreated through algebraic progression by either one of

- $U(n) = U(n - F(n) + U(n + F(n-2))$ and
- $U(n) = U(n - F(n) - U(n + F(n+2))$.

147

An Aether expressing Fibonacci and Lucas Sequences joined at the hip: a thought experiment

The generation of harmonic balanced sequences and unbalanced sequences using a fundamental Fibonacci sequence as a baseline.

Up to now I´ve been treating universal algebraic sequences as is there is only one example of an Aether-based Fibonacci sequence and Lucas sequence to be achieved algebraically. In essence it comes down to achieving values for variable of U(1) and U(-1) on each side of a nexus point whereby F(1) = F(-1).

As a precondition for the generation of algebraic harmonic variations of the Fibonacci and Lucas sequences, it is taken a priori that:

- a) Aether "states" already exist as a baseline; these states being generated through a process following the expression of EQN 5

$$F(n) = F\left(\frac{(2n + 3 + i^{2n})}{4}\right)^2 - i^{2n}F\left(\frac{(2n - 3 - i^{2n})}{4}\right)^2$$

where "i" = square root of -1 (- i.e. $\sqrt{-1}$),

[

EQN 5]

and that at a minimum

- b) these Aether-"states" may interact with one another either through addition or subtraction to

An Aether expressing Fibonacci and Lucas Sequences joined at the hip: a thought experiment

produce either a sum-"state" or a difference-"state" respectively.

Initial premise for the generation of a harmonic sequence

As mentioned earlier, it is assumed that Aether-based states already exist through expansion from a nexus point F(0) and employing EQN 5. Thus "states"

n	Fibonacci Base Line
-7	13
-6	-8
-5	5
-4	-3
-3	2
-2	-1
-1	1
0	0
1	1
2	1
3	2
4	3
5	5
6	8
7	13

of complementary "time-strand" iterations pairs for "n" of "0" to "-7" and "0" to "7" are considered prime facie to be present or pre-existing in an Aether-space.

An Aether expressing Fibonacci and Lucas Sequences joined at the hip: a thought experiment

For creating the harmonic algebraic sequences, much like the interrelated sequences of the Fibonacci sequence and Lucas sequence as we generally know them, the sequences $U_S(n+)$ and $U_S(n-)$ will be generated through expressions:

- $U_S(n+) = F(n+1) + F(n-a)$; and
- $U_S(n-) = F(n+1) - F(n-a)$.

where the variable "a" is a positive integer greater than or equal to "1".

Examples of harmonic sequence building

For example, in the case of "a" = 1, we then have the expressions:

- $U_S(n+) = F(n+1) + F(n-1)$; and
- $U_S(n-) = F(n+1) - F(n-1)$,

which are expressions which we have already shown to give values which are identical to those of the Fibonacci sequence and Lucas sequence respectively.

In this respect see:

An Aether expressing Fibonacci and Lucas Sequences joined at the hip: a thought experiment

F(n) = F(n+1) - F(n-1)		L(n) = F(n+1) + F(n-1)	
34		-76	
-21		47	
13		-29	
-8		18	
5		-11	
-3		7	
2		-4	
-1		3	
1		-1	
0		2	
1		1	
1		3	
2		4	
3		7	
5		11	
8		18	
13		29	
21		47	
34		76	
Cassini Idenities	1		-5
Average of Cassini Values	3	L(2) or F(4)	
Diffence in Cassini Values		4	

As can be seen here, I have reproduced the Fibonacci and Lucas sequences for "a" = 1. Also included are some characteristics related to the Cassini Identity.

151

An Aether expressing Fibonacci and Lucas Sequences joined at the hip: a thought experiment

If we progress further with increasing values for "a" with "a" equal to 2 and 3 we get the following:

U(n-) = F(n+1) - F(n-2)	U(n+) = F(n+1) + F(n-2)		U(n-) = F(n+1) - F(n-3)	U(n+) = F(n+1) + F(n-3)	
68	68		-76	102	
-42	-42		47	-63	
26	26		-29	39	
-16	-16		18	-24	
10	10		-11	15	
-6	-6		7	-9	
4	4		-4	6	
-2	-2		3	-3	
2	2		-1	3	
0	0		2	0	
2	2		1	3	
2	2		3	3	
4	4		4	6	
6	6		7	9	
10	10		11	15	
16	16		18	24	
26	26		29	39	
42	42		47	63	
68	68		76	102	
Cassini Idenities	4	4	Cassini Idenities	-5	9
Average of Cassini Values	4	L(3)	Average of Cassini Values	7	L(4)
Diffence in Cassini Values	0		Diffence in Cassini Values	4	

What is noticeable here is that for "a" = 2, it does not matter whether the variable "F(n-2)" is added or subtracted, the result is the same. Both expressions for U(n) give a sequence of values which are simply multiples of "2" of the base line Fibonacci sequence. In this respect, it is submitted that this outcome in terms of expected base units may illustrate possible shortcomings in experiments

152

An Aether expressing Fibonacci and Lucas Sequences joined at the hip: a thought experiment

in physics where the outcome thereof is dependent on a material whose component under examination (e.g., an electron) is expected to obey an inverse square law (i.e., the converse of the expression of EQN 5 used to generate the base line sequence). That is, if for some reason a base unit is assumed to comprise a single active component only, could one distinguish the outcome of that experiment from that in which the base unit of the component under examination is in fact comprised of two identical and complementary components (e.g., mutually tidal locked electrons).

If we continue further to "a" = 3, we see the outcome again giving rise to complementary Fibonacci and Lucas sequences, with the "F(n-3)" of the U(n-) sequence giving rise to values identical to that of the Lucas sequence, but with the "F(n-3)" of the U(n+) sequence giving rise to values identical to that of the Fibonacci sequence having a base unit of "3" times [F(4)] that of the "base line" sequence.

Thus, in systems involving $F(4)^2$, as per the Rodin-type Force graph, do the alleged vortices thereof arise merely out of a simple algebraic sum involving the states of the base line Fibonacci sequence? In this case for "a" = 3, the Cassini identity for U(n+)

-    $F(0)^2 + F(-1) * F(1)$

rather than being equal to "1" as per the base line Fibonacci sequence, in fact becomes "9" as both F(-1) and F(1) have values of "3" [F(4)] each.

An Aether expressing Fibonacci and Lucas Sequences joined at the hip: a thought experiment

For full disclosure, it is admitted that in constructing these tables for the harmonics of the Fibonacci and Lucas sequences, the presentation of these tables has been manipulated to a degree. But only in so far that the sequences have been shifted relative to one another so that each of the "inflection points" for each sequence are coincident on the same row. Otherwise, all the values are derived from simple subtraction or addition between values of the "base line" Fibonacci sequence.

Four further examples of harmonic sequences shall now be presented.

U(n-) = F(n+1) - F(n-4)	U(n+) = F(n+1) + F(n-4)		U(n-) = F(n+1) - F(n-5)	U(n+) = F(n+1) + F(n-5)	
157	81		136	-152	
-97	-50		-84	94	
60	31		52	-58	
-37	-19		-32	36	
23	12		20	-22	
-14	-7		-12	14	
9	5		8	-8	
-5	-2		-4	6	
4	3		4	-2	
-1	1		0	4	
3	4		4	2	
2	5		4	6	
5	9		8	8	
7	14		12	14	
12	23		20	22	
19	37		32	36	
31	60		52	58	
50	97		84	94	
81	157		136	152	
Cassini Idenities	11	11	Cassini Idenities	16	-20
Average of Cassini Values	11	L(5)	Average of Cassini Values	18	L(6)
Diffence in Cassini Values	0		Diffence in Cassini Values	4	

Tables for complementary sequence pairs for U(n)= F(n+1) – F(n- a) and U(n) = F(n+1) + F(n- a) for a = 4 and a = 5

The values for sequences U(n-) and U(n+) generated from the variable F(n-4) differ from the three previous sequences in that they are each "unbalanced". In this respect, each sequence for "a" = 4 has an inferred

155

An Aether expressing Fibonacci and Lucas Sequences joined at the hip: a thought experiment

inflection point at "-1" and "1" respectively. The sequences are however nevertheless mutual inversions of one another in that the absolute values when read in up-down in one sequence are the converse of the values when read down-up in the other sequence. In addition, it is noted that where there appears to be a "reactance-dominated" portion of the sequence in one is the "real-dominated" portion of the sequence in the other [the reactance dominated portion of a sequence being that portion where positive-negative oscillation occurs].

The values for sequences $U(n-)$ and $U(n+)$ generated from the variable "$F(n-5)$" are again respective "balanced sequences", $U(n-)$ again taking Fibonacci characteristics and $U(n+)$ taking Lucas characteristics. Whereas the $U(n-)$ sequence appears to be a Fibonacci harmonic modified through multiplication by $L(3)$ [=4], the $U(n+)$ sequence appears to be a Lucas harmonic modified through multiplication by $F(3)$[=2]. In this case there is a clear relation between the multipliers from which these harmonics are formed.

An Aether expressing Fibonacci and Lucas Sequences joined at the hip: a thought experiment

$U(n-) =$ $F(n+1) - F(n-6)$	$U(n+) =$ $F(n+1) + F(n-6)$	$U(n-) =$ $F(n+1) - F(n-7)$	$U(n+) =$ $F(n+1) + F(n-7)$
149	225	-228	238
-92	-139	141	-147
57	86	-87	91
-35	-53	54	-56
22	33	-33	35
-13	-20	21	-21
9	13	-12	14
-4	-7	9	-7
5	6	-3	7
1	-1	6	0
6	5	3	7
7	4	9	7
13	9	12	14
20	13	21	21
33	22	33	35
53	35	54	56
86	57	87	91
139	92	141	147
225	149	228	238

Cassini Idenities	29		29	Cassini Idenities	-45		49
Average of Cassini Values	29	L(7)		Average of Cassini Values	47	L(8)	
Diffence in Cassini Values		0		Diffence in Cassini Values		4	

If we now turn to sequences generated for variables "F(n-6)" and "F(n-7)", we see the latter story of two mutually complemented unbalanced sequences then followed by two balanced sequences being repeated.

157

An Aether expressing Fibonacci and Lucas Sequences joined at the hip: a thought experiment

Tables for "a" = 1 to 20 for harmonic pair sequences

There are more harmonic sequences to be presented, but for efficiency, the following tables of characteristics is provided.

"a"	"-" / "+"	Balanced/ Unbalanced B/U	Fibonacci (F) / Lucas (L) seq. times F(x)/L(x)	Cassini Value (Absolute)	Cassini Value L(x) -/+ F(0)/L(0)
			Characteristics of paired sequences U(n) arising from expressions U(n-) = F(n+1) - F(n-a) and U(n+) = F(n+1) + F(n-a)		
1	"-"	B	F x L(1)	1	L(2) - L(0)
	"+"		L x F(1)	5 (-5)	L(2) + L(0)
2	"-"	B	F x L(0)	4	*L(3) - F(0)*
	"+"		F x L(0)	4	*L(3) + F(0)*
3	"-"	B	L x F(2)	5 (-5)	L(4) - L(0)
	"+"		F x L(2)	9	L(4) + L(0)
4	"-"	U	U1-	11	*L(5) - F(0)*
	"+"		U1+	11	*L(5) + F(0)*
5	"-"	B	F x L(3)	16	L(6) - L(0)
	"+"		L x F(3)	20(-20)	L(6) + L(0)
6	"-"	U	U2-	29	*L(7) - F(0)*
	"+"		U2+	29	*L(7) + F(0)*
7	"-"	B	L x F(4)	45(-45)	L(8) - L(0)
	"+"		F x L(4)	49	L(8) + L(0)
8	"-"	U	U3-	76	*L(9) - F(0)*
	"+"		U3+	76	*L(9) + F(0)*
9	"-"	B	F x L(5)	121	L(10) - L(0)
	"+"		L x F(5)	125(-125)	L(10) + L(0)
10	"-"	U	U4-	199	*L(11) - F(0)*
	"+"		U4+	199	*L(11) + F(0)*

And again, with the balanced sequences, we see the same mutual relationship where the Lucas-based harmonic is

An Aether expressing Fibonacci and Lucas Sequences joined at the hip: a thought experiment

multiplied by F(4) [= 3] and the Fibonacci-based harmonic is multiplied by L(4) [=7].

As can be noted from the above, except for "a" = 2 all complementary pairs of even valued a´s are unbalanced.

"a"	"-" / "+"	Balanced/ Unbalanced B/U	Fibonacci (F) / Lucas (L) seq. times F(x)/L(x)	Cassini Value (Absolute)	Cassini Value L(x) -/+ F(0)/L(0)
			Characteristics of paired sequences U(n) arising from expressions U(n-) = F(n+1) - F(n-a) and U(n+) = F(n+1) + F(n-a)		
11	"-"	B	L x F(6)	320(-320)	L(12) - L(0)
	"+"		F x L(6)	324	L(12) + L(0)
12	"-"	U	U5-	521	L(13) - F(0)
	"+"		U5+	521	L(13) + F(0)
13	"-"	B	F x L(7)	841	L(14) - L(0)
	"+"		L x F(7)	845(-845)	L(14) + L(0)
14	"-"	U	U6-	1364	L(15) - F(0)
	"+"		U6+	1364	L(15) + F(0)
15	"-"	B	L x F(8)	2205(-2205)	L(16) - L(0)
	"+"		F x L(8)	2209	L(16) + L(0)
16	"-"	U	U7-	3571	L(17) - F(0)
	"+"		U7+	3571	L(17) + F(0)
17	"-"	B	F x L(9)	5776	L(18) - L(0)
	"+"		L x F(9)	5780(-5780)	L(18) + L(0)
18	"-"	U	U8-	9349	L(19) - F(0)
	"+"		U8+	9349	L(19) + F(0)
19	"-"	B	L x F(10)	15125(-15125)	L(20) - L(0)
	"+"		F x L(10)	15129	L(20) + L(0)
20	"-"	U	U9-	24476	L(21) - F(0)
	"+"		U9+	24476	L(21) + F(0)

An Aether expressing Fibonacci and Lucas Sequences joined at the hip: a thought experiment

In contrast all pairs of *odd*-valued "a´s" comprise a complementary harmonic of Fibonacci-type sequence and a Lucas-type sequence. The pair for "a" = 2 is also exceptional in that it comprises only the same identical sequence for each sequence of iterations.

Relationship between "a" and mutually complementary Fibonacci and Lucas element multipliers

As also illustrated by the tables here, the Cassini Identities for each of the sequences themselves follows in essence a sequence of Lucas values with the relation $CI_a = L(1 + a)$. In the case of "a" associated with unbalanced sequences, there is only one CI_a value shared by both sequences. In the case of pairs of balanced sequences, the Cassini Identity differs from CI_a by either subtracting $L(0)$ or adding $L(0)$.

The odd (unbalanced) pairs differ from a respective central Lucas value by a value of "2" [$L(0)$] if only considering the absolute values of CI_a. If one considers that the polarities of each of the CI_a values are also different, then it appears that these CI_a values on mutually opposite sides of an inflection point of $F(0)$.

The embedded Catalan identity within the paired sequences

It is further noted that for each of the modified CI_a Lucas values of balanced sequences which have a negative polarity, these values are:

An Aether expressing Fibonacci and Lucas Sequences joined at the hip: a thought experiment

- multiples of "5"; and
- when each modified Lucas value is divided by "5", the resulting values follow a progression with values of "a" of

$$(+)1\_ (-)1\_ (+)4\_ (-)9\_ (+)25\_ (-)64\_ \ldots$$

wherein (-) or (+) indicate whether the harmonic sequence associated with the value "a", results from the element "F(n-a)" being added or subtracted from the element "F(n+1)".

As such therefor the Catalan Identity is itself embedded in the progression of paired sequences. Fractals within fractals. Its fractals all the way down.

An Aether expressing Fibonacci and Lucas Sequences joined at the hip: a thought experiment

Bits, words, and Syntax of Modulo Number System (MNS) circuits

As previously discussed, regarding MNS-F(n)^2 circuits when reproducing the Fibonacci or Lucas sequence, these sequences form closed loops. What has not been discussed yet, is how these sequences are give rise to values which are square roots of (or integer multiples of) the Modulo value.

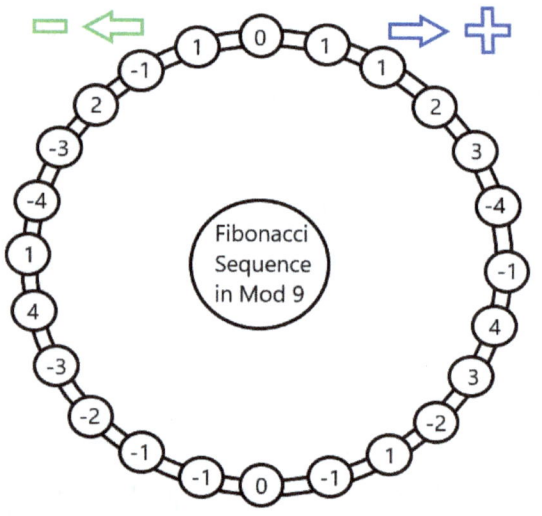

S

pin corrected 24-element Fibonacci sequence cycle in MNS-9 circuit. To the left, F(-(n+1) = F(-(n-1) – F(-n) and to the right, F(n+1) = F(n-1) + F(n).

162

An Aether expressing Fibonacci and Lucas Sequences joined at the hip: a thought experiment

That is, taking the modulo value as being the equivalent of zero, how does one treat the values in the sequence which are in essence acting roots of zero.

Tables and time-strands

Let's look at a sequence table for the 24-element MNS-$F(4)^2$ circuit with modulo-9. To read the tables, one must consider:

- "0" represents the inflection point of the Ever Present Now (EPN), to the left extends the forward going time string and to the right the backward going time strand;
- $F(x)$ references values of the forward going time strand;
- $F(-x)$ references values of the backward going time strand; and
- $F(x) * F(-x)$ is the product of the forward and backward going time strands interacting with one another to form a square value.

The strike marks to the right indicate instances where either a "0" has occurred (two marks), or where a root of "0" has occurred (single mark).

An Aether expressing Fibonacci and Lucas Sequences joined at the hip: a thought experiment

Standard				n	Spin Corrected				
F(x)		F(-x)	F(x) * F(-x)		F(x)		F(-x)	F(x) * F(-x)	
	0		0	0		0		0	
1		1	1	1	1		1	1	
1		-1	-1	2	1		-1	-1	
2		2	4	3	2		2	4	
3		-3	0	4	3		-3	0	I
5		5	7	5	-4		-4	-2	
8		-8	8	6	-1		1	-1	
4		4	7	7	4		4	-2	
3		-3	0	8	3		-3	0	I
7		7	4	9	-2		-2	4	
1		-1	-1	10	1		-1	-1	
8		8	1	11	-1		-1	1	
0		0	0	12	0		0	0	I I
8		8	1	13	-1		-1	1	
8		-8	8	14	-1		1	-1	
7		7	4	15	-2		-2	4	
6		-6	0	16	-3		3	0	I
4		4	7	17	4		4	-2	
1		-1	-1	18	1		-1	-1	
5		5	7	19	-4		-4	-2	
6		-6	0	20	-3		3	0	I
2		2	4	21	2		2	4	
8		-8	8	22	-1		1	-1	
1		1	1	23	1		1	1	
0		0	0	24	0		0	0	I I

24 elements of forward and backward time strings and the products thereof in an MNS-9 circuit

As can be seen from the table above, with every fourth iteration, the product of F(x) and F(-x) is zero. This results either from

- each element F(x) and F(-x) being equal to zero [two strike marks to the right], or

An Aether expressing Fibonacci and Lucas Sequences joined at the hip: a thought experiment

- each element being equal to "3" (i.e., the square root of "9" which, for the MNS-9 circuit, is equivalent to "0") [one strike mark to the right].

One must also note, especially in the "spin corrected" sequence for "F(x) * F(-x)", that there are two identical palindromes formed extending from n = 0 to 12 and from n = 12 to 24. As such therefor I submit that the product of F(x) * F(-x) provide a basis for the generation of standing waves within the Aether.

Using "0" `s and "4"'s to establish bits and words of a syntax for time strand products in Fibonacci sequence circuits and Lucas sequence circuits.

To the right, based on the MNS-F(4)^2 circuit as an example, a sequence of the 24 products of F(x) * F(-x) for the Fibonacci sequence is illustrated. Taking a cursory look, one item which pops out is that the sequence is composed of two identical palindromes in series. What is also noted is that every fourth element comprises a "0".

It is proposed here for F(x)^2 in general that if we:

- take each non-zero element of the product sequence F(x) x F(-x) as being the equivalent of a "bit" in a bitstream; and

165

An Aether expressing Fibonacci and Lucas Sequences joined at the hip: a thought experiment

- take each "0"-element of the sequence as being a means of punctuating the bitstream of the sequence,

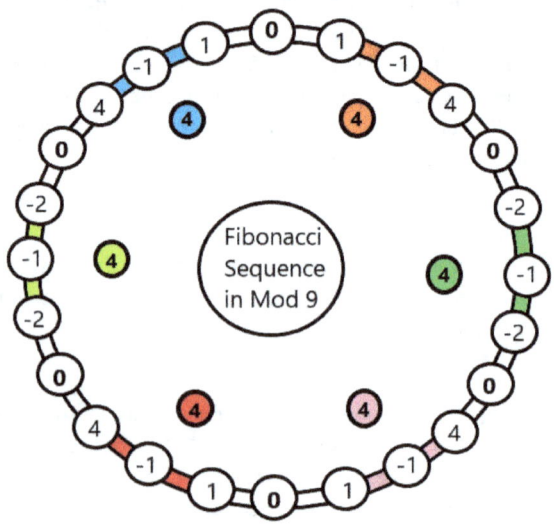

Generation of carrier states of Aether-value "4" from sum of squares of elements of Fibonacci sequence using "0"as punctuation

then we have a process of generating "words" of length "x-1" in an MNS-F(x)^2 circuit. In this case of "x" being equal to "4", each word of the MNS-9 "product" sequence has a spin-corrected value of "4".

An Aether expressing Fibonacci and Lucas Sequences joined at the hip: a thought experiment

Now, based on MNS-F(4)^2, is a sequence of products of L(x) x L(-x) for the Lucas sequence.

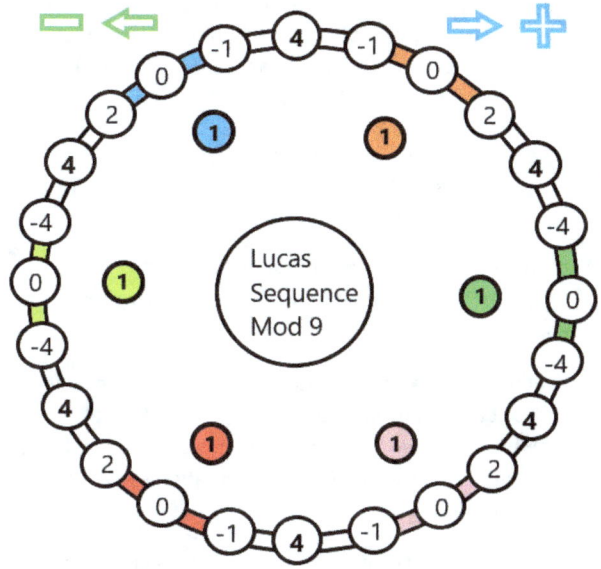

Generation of carrier states of Aether-value "1" from sum of squares of elements of Lucas sequence using "4" as punctuation (where "4" corresponds to "0" of the Fibonacci sequence)

Taking again a cursory look, this sequence is also composed of two identical palindromes in series. However, what is also noted is that every fourth element, in contrast to the

An Aether expressing Fibonacci and Lucas Sequences joined at the hip: a thought experiment

products of the Fibonacci circuit for MNS-9, comprises a "4"-state instead of a "0"-state, in essence the "2"-operator squared. It is proposed here therefor that if we:

- take each non-four element of the product sequence L(x) * L(-x) as being the equivalent to a "bit"; and

- take each "4"-element of the sequence as being a means of punctuating the "bit"-stream of the sequence,

then we have a second process of generating "words" of length "x-1". In the case of MNS-9 of the Lucas sequence, each "word" of the "product" has a "1"-state.

In addition, what is also noted from the MNS-9 circuit is that at points shifted by two "bits", there exists also a "0"-bit which is repeated at every fourth element. Assuming we can still take the value "0" for punctuation even in this case, then again lets:

- take each non-"0"-element of the product sequence L(x) * L(-x) between the "0"-states as being the equivalent to a "bit"; and

- take each "0"-element of the sequence as being a means of punctuating the bitstream of the sequence,

and we have, for the Lucas sequence, a further second process of generating "words" of length "x-1".

An Aether expressing Fibonacci and Lucas Sequences joined at the hip: a thought experiment

In this case of MNS-F(4)^2, each word of the "product" sequence of the MNS-F(4)^2 circuit has a value of "2".

At this point we can see that the Fibonacci and complementary Lucas sequences of MNS-F(4)^2 continuously generate three words which are repeated:

- a superposition of two word-states in parallel of a "4"-state word of the Fibonacci sequence and a "1"-state word of the complementary Lucas sequence; and
- a "2"-state word of the complementary Lucas sequence intermediates the two previous words.

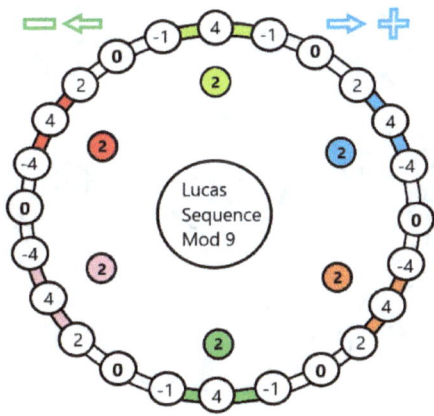

Generation of carrier states of Aether-value "2" from sum of squares of elements of Lucas sequence using "0" as punctuation

169

An Aether expressing Fibonacci and Lucas Sequences joined at the hip: a thought experiment

Using "0" `s and root "0" ´s to establish bits and words of a syntax for time strands in Fibonacci sequence circuits and Lucas sequence circuits.

Up to now we have been dealing with sequences of products. Now we shall deal with the sequence of elements F(x) and L(x). Different from the "words" generated from "products", not only will:

- "0" be used to demarcate punctuation,

but also the root of "0" and multiples thereof.

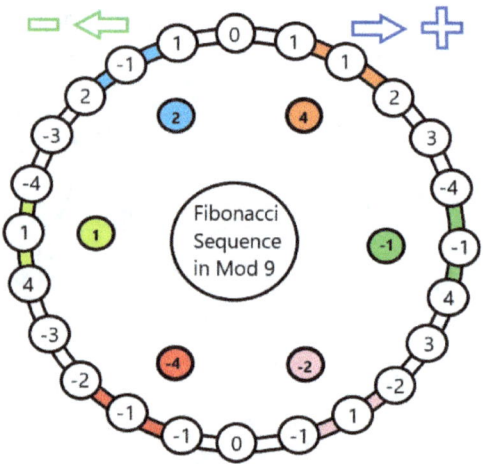

Spin corrected Fibonacci sequence with forward clockwise time strand and reverse anti-clockwise time strand directions.

170

An Aether expressing Fibonacci and Lucas Sequences joined at the hip: a thought experiment

Thus, in the case of MNS-9, the values „3" and „-3" are treated as having the properties of dualling numbers and are used to demarcate where a "word" of each time "time strand" begins and ends. As can be seen, using these rules of demarcation, we end up with a repeating sequence of "words" of

-   1 _ 2 _ 4 _ - 1 _ - 2 _ - 4

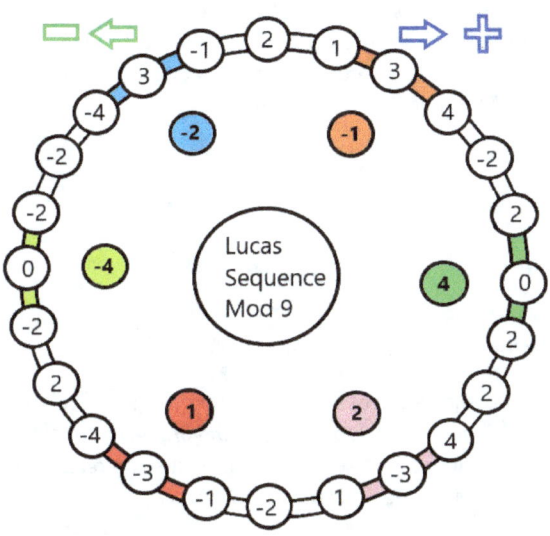

Spin corrected Lucas sequence with forward clockwise time strand and reverse anti-clockwise time strand directions.

An Aether expressing Fibonacci and Lucas Sequences joined at the hip: a thought experiment

In this case we have the spin corrected Lucas Sequence in MNS-9. But now, rather than using values of "0" to act as demarcation for words, the values of "2" and "-2" corresponding to "4" in the "words" of products are used. Thus now:

- Pairs of "2" and "-2" are to be used to demarcate punctuation of the "words".

As can be seen, using these rules of demarcation, we end up with a sequence like that of the Fibonacci sequence, but which is the reverse order of the repeating sequence of "words" thereof. That is, a sequence:

- 4 _ 2 _ 1 _ - 4 _ - 2 _ - 1.

Terminology regarding words and states in Aether circuits

Now for some terminology. In these cases:

- the values of each of the "words" generated from demarcation of *"time strand products"* in Fibonacci and Lucas MNS circuits are referred to as *"carrier-states"*; and

- the values of each of the "words" generated from demarcation of *"time strand elements"* in Fibonacci and Lucas systems are referred to as *"Supra-MNS-states"*.

An Aether expressing Fibonacci and Lucas Sequences joined at the hip: a thought experiment

Examples of Carrier states and Supra-MNS states for systems comprising MNS-F(x)^2 and MNS-L(x)^2 circuits

MNS-F(3)^2 circuit with modulo "4"

As mentioned earlier, MNS states involving each of MNS-F(0)^2, MNS-F(1)^2 and MNS-F(2)^2 cannot be determined as, per definition of a modulo system, each element of these systems would only return zero. That is:

- for MNS-F(0)^2, anything equal to or greater than "0" would return a value of "0"; and
- for each of MNS-F(1)^2 and MNS-F(2)^2, anything equal to or greater than "1" would return a value of "0".

Admittedly, we can divide "0" by "0" to obtain "2" or multiply "2" by "0" to get "0", but as of yet we do not have a definition of what result may be achieved through multiplying "0" by "0" (other than the presumption that it is simply "0"). This then leaves MNS-F(3)^2 as the smallest MNS-circuit for which we can obtain values for Carrier-states and Supra-MNS states.

An Aether expressing Fibonacci and Lucas Sequences joined at the hip: a thought experiment

Tables for Fibonacci sequence with MNS-F(3)² circuit

Standard				n	Spin Corrected		
F(x)		F(-x)	F(x) * F(-x)		F(x)	F(-x)	F(x) * F(-x)
	0		0	0	0		0
1		1	1	1	1	1	1
1		-1	-1	2	1	-1	-1
2		2	0	3	2	2	0
3		-3	3	4	-1	1	-1
1		1	1	5	1	1	1
0		0	0	6	0	0	0

Carrier States			Word Index n	Supra-MNS states		
Original Sum		Corrected Sum		Sum (x+)		Sum (x-)
	0				0	
0		0	1	2		0
0		0	2	0		2

It must be noted here that the value of the "carrier state" here is indicated as "0" only. However as "4" in the MNS-4 system is also indicated as being "0", this value of "0" may also be considered synonymous with a value of "4". Reasons for this clarification will become clear later.

An Aether expressing Fibonacci and Lucas Sequences joined at the hip: a thought experiment

Tables for Fibonacci sequence with MNS-F(4)^2 circuit

Standard			n	Spin Corrected			
F(x)	F(-x)	F(x) * F(-x)	n	F(x)	F(-x)	F(x) * F(-x)	
0		0	0	0		0	
1	1	1	1	1	1	1	
1	-1	-1	2	1	-1	-1	
2	2	4	3	2	2	4	
3	-3	0	4	3	-3	0	I
5	5	7	5	-4	-4	-2	
8	-8	8	6	-1	1	-1	
4	4	7	7	4	4	-2	
3	-3	0	8	3	-3	0	I
7	7	4	9	-2	-2	4	
1	-1	-1	10	1	-1	-1	
8	8	1	11	-1	-1	1	
0	0	0	12	0	0	0	I I
8	8	1	13	-1	-1	1	
8	-8	8	14	-1	1	-1	
7	7	4	15	-2	-2	4	
6	-6	0	16	-3	3	0	I
4	4	7	17	4	4	-2	
1	-1	-1	18	1	-1	-1	
5	5	7	19	-4	-4	-2	
6	-6	0	20	-3	3	0	I
2	2	4	21	2	2	4	
8	-8	8	22	-1	1	-1	
1	1	1	23	1	1	1	
0	0	0	24	0	0	0	I I

An Aether expressing Fibonacci and Lucas Sequences joined at the hip: a thought experiment

Carrier States				Supra-MNS states		
Original Sum		Corrected Sum	Word Index n	Sum (x+)		Sum (x-)
	0				0	
4		4	1	4		2
-5		4	2	8		1
4		4	3	7		5
4		4	4	5		7
-5		4	5	1		8
4		4	6	2		4

For clarification, I am treating "0" as the inflection point between forward time and reverse time, and not as part of the sequence per se. One should therefor note that between the two sequences of super-MNS states separated by the "0", that there is a continuity in the Supra-MNS states as one moves down the words of each of the forward times strands (Sum(x+)) and the reverse time strands (Sum(x-)). That is both sequences of words behave as if they are part of the same sequence words. This could be analogous to the passing of time with past and future strands extending individually e.g., from the consciousness that is the "0" of the Ever Present Now (EPN).

An Aether expressing Fibonacci and Lucas Sequences joined at the hip: a thought experiment

Tables for Fibonacci sequence with MNS-F(5)² circuit

F(x)		F(-x)	F(x) * F(-x)	n	F(x)		F(-x)	F(x) * F(-x)	
	0		0	0		0		0	
1		1	1	1	1		1	1	
1		-1	-1	2	1		-1	-1	
2		2	4	3	2		2	4	
3		-3	-9	4	3		-3	-9	
5		5	0	5	5		5	0	I
8		-8	11	6	8		-8	11	
13		13	19	7	-12		-12	-6	
21		-21	9	8	-4		4	9	
9		9	6	9	9		9	6	
5		-5	0	10	5		-5	0	I
14		14	21	11	-11		-11	-4	
19		-19	14	12	-6		6	-11	
8		8	14	13	8		8	-11	
2		-2	-4	14	2		-2	-4	
10		10	0	15	10		10	0	I
12		-12	6	16	12		-12	6	
22		22	9	17	-3		-3	9	
9		-9	19	18	9		-9	-6	
6		6	11	19	6		6	11	
15		-15	0	20	-10		10	0	I
21		21	16	21	-4		-4	-9	
11		-11	4	22	11		-11	4	
7		7	24	23	7		7	-1	
18		-18	1	24	-7		7	1	
0		0	0	25	0		0	0	I I

The Fibonacci sequence with the MNS-25 circuit in fact requires 100 iterations before the cycle begins again. For efficiency, only the first 25 are shown here.

An Aether expressing Fibonacci and Lucas Sequences joined at the hip: a thought experiment

Carrier States				Supra-MNS states		
Original Sum		Corrected Sum	Word Index n	Sum (x+)		Sum (x-)
	0				0	
-5		-5	1	7		24
20		-5	2	1		18
-30		-5	3	18		1
20		-5	4	24		7
-5		-5	5	7		24
-5		-5	6	1		18
20		-5	7	18		1
-30		-5	8	24		7
20		-5	9	7		24
-5		-5	10	1		18
-5		-5	11	18		1
20		-5	12	24		7
-30		-5	13	7		24
20		-5	14	1		18
-5		-5	15	18		1
-5		-5	16	24		7
20		-5	17	7		24
-30		-5	18	1		18
20		-5	19	18		1
-5		-5	20	24		7

Carrier states and Supra-MNS states for each of the 20 words from the 100 iterations of a full cycle.

An Aether expressing Fibonacci and Lucas Sequences joined at the hip: a thought experiment

Summary of states for MNS-F(4)2 and MNS-F(5)^2

If we compare the Supra-MNS states for MNS-F(4)^2 with those of MNS-F(5)^2 we note:

- whereas the Supra-MNS states of MNS-F(4)^2 appear to be generated through multiplication by the Fibonacci Cassini Identity multipliers F(3) [=2] and F(5) [=5],

the Supra-MNS states of MNS-F(5)^2 appear to be generated through multiplication by the corresponding Lucas Cassini Identity multipliers L(4) [=7] and L(6) [=18 (or spin corrected -7)].

It is also noteworthy in this respect that the respective sums of groups of the supra-MNS states of the MNS-F(4)^2 circuit (which amount to 7 (= 1 + 2 + 4) and -7 (= -1 - 2 - 4) respectively), equals the values of each of the complementary multipliers for the Supra-MNS states of the MNS-F(5)^2 circuit.

About the Carrier states of each of the MNS-F(4)^2 and MNS-F(5)^2 it is noted that whereas the carrier state for MNS-F(4)^2 switches between "4" and "-5" when not spin-corrected, the carrier state for MNS-F(5)^2 switches between "-5" [i.e., "-30" and "-5"] and "20".

179

An Aether expressing Fibonacci and Lucas Sequences joined at the hip: a thought experiment

Tables for Fibonacci sequence with MNS-F(6)² circuit

Standard			n	Spin Corrected			
F(x)	F(-x)	F(x) * F(-x)		F(x)	F(-x)	F(x) * F(-x)	
0		0	0	0		0	
1	1	1	1	1	1	1	
1	-1	-1	2	1	-1	-1	
2	2	4	3	2	2	4	
3	-3	-9	4	3	-3	-9	
5	5	25	5	5	5	25	
8	-8	0	6	8	-8	0	I
13	13	41	7	13	13	-23	
21	-21	7	8	21	-21	7	
34	34	4	9	-30	-30	4	
55	-55	47	10	-9	9	-17	
25	25	49	11	25	25	-15	
16	-16	0	12	16	-16	0	I
41	41	17	13	-23	-23	17	
57	-57	15	14	-7	7	15	
34	34	4	15	-30	-30	4	
27	-27	39	16	27	-27	-25	
61	61	9	17	-3	-3	9	
24	-24	0	18	24	-24	0	I
21	21	57	19	21	21	-7	
45	-45	23	20	-19	19	23	
2	2	4	21	2	2	4	
47	-47	31	22	-17	17	31	
49	49	33	23	-15	-15	-31	
32	-32	0	24	32	-32	0	I

The Fibonacci sequence with the MNS-64 circuit requires 96 iterations before the cycle begins again. For efficiency, only the first 24 are shown here.

An Aether expressing Fibonacci and Lucas Sequences joined at the hip: a thought experiment

Carrier States				Supra-MNS states		
Original Sum		Corrected Sum	Word Index n	Sum (x+)		Sum (x-)
	0				0	
20		20	1	12		4
-44		20	2	20		60
20		20	3	28		52
20		20	4	36		44
20		20	5	44		36
20		20	6	52		28
-44		20	7	60		20
20		20	8	4		12
20		20	9	12		4
-44		20	10	20		60
20		20	11	28		52
20		20	12	36		44
20		20	13	44		36
20		20	14	52		28
-44		20	15	60		20
20		20	16	4		12

Carrier states and Supra-MNS states for each of the 16 words from the 96 iterations of a full cycle.

- Non-spin corrected carrier states oscillate between "20" and "-44"; and
- Supra-MNS states follow an algebraic progression of $U(n+1) = U(n) + F(6)$ where $F(0) = $ "4" i.e., $L(0)^2$.

An Aether expressing Fibonacci and Lucas Sequences joined at the hip: a thought experiment

Tables for Fibonacci sequence with MNS-F(7)² circuit

Standard			n	Spin Corrected			
F(x)	F(-x)	F(x) * F(-x)	n	F(x)	F(-x)	F(x) * F(-x)	
0		0	0	0		0	
1	1	1	1	1	1	1	
1	-1	-1	2	1	-1	-1	
2	2	4	3	2	2	4	
3	-3	-9	4	3	-3	-9	
5	5	25	5	5	5	25	
8	-8	-64	6	8	-8	-64	
13	13	0	7	13	13	0	I
21	-21	66	8	21	-21	66	
34	34	142	9	34	34	-27	
55	-55	17	10	55	-55	17	
89	89	147	11	-80	-80	-22	
144	-144	51	12	-25	25	51	
64	64	40	13	64	64	40	
39	-39	0	14	39	-39	0	I
103	103	131	15	-66	-66	-38	
142	-142	116	16	-27	27	-53	
76	76	30	17	76	76	30	
49	-49	134	18	49	-49	-35	
125	125	77	19	-44	-44	77	
5	-5	-25	20	5	-5	-25	
130	130	0	21	-39	-39	0	I
135	-135	27	22	-34	34	27	
96	96	90	23	-73	-73	-79	
62	-62	43	24	62	-62	43	
158	158	121	25	-11	-11	-48	
51	-51	103	26	51	-51	-66	
40	40	79	27	40	40	79	
91	-91	0	28	-78	78	0	I

The Fibonacci sequence with the MNS-169 circuit requires 364 iterations before the cycle begins again. For efficiency, only the first 28 are shown here.

An Aether expressing Fibonacci and Lucas Sequences joined at the hip: a thought experiment

Carrier States				Supra-MNS states		
Original Sum		Corrected Sum	Word Index n	Sum (x+)		Sum (x-)
	0				0	
-44		-44	1	20		165
125		-44	2	69		136
-44		-44	3	162		108
-44		-44	4	35		46
-44		-44	5	163		126
-44		-44	6	30		110
294		-44	7	19		147
-44		-44	8	74		72
-44		-44	9	137		87
-44		-44	10	160		84
-44		-44	11	45		17
125		-44	12	113		98
-44		-44	13	111		48
-44		-44	14	121		58
125		-44	15	71		56
-44		-44	16	152		124
-44		-44	17	85		9
-44		-44	18	82		32
-44		-44	19	97		95
294		-44	20	22		150
-44		-44	21	59		139
-44		-44	22	43		6
-44		-44	23	123		134
-44		-44	24	61		7
125		-44	25	33		100
-44		-44	26	4		149

An Aether expressing Fibonacci and Lucas Sequences joined at the hip: a thought experiment

Carrier States				Supra-MNS states		
Original Sum		Corrected Sum	Word Index n	Sum (x+)		Sum (x-)
	0				0	
-44		-44	27	149		4
125		-44	28	100		33
-44		-44	29	7		61
-44		-44	30	134		123
-44		-44	31	6		43
-44		-44	32	139		59
294		-44	33	150		22
-44		-44	34	95		97
-44		-44	35	32		82
-44		-44	36	9		85
-44		-44	37	124		152
125		-44	38	56		71
-44		-44	39	58		121
-44		-44	40	48		111
125		-44	41	98		113
-44		-44	42	17		45
-44		-44	43	84		160
-44		-44	44	87		137
-44		-44	45	72		74
294		-44	46	147		19
-44		-44	47	110		30
-44		-44	48	126		163
-44		-44	49	46		35
-44		-44	50	108		162
125		-44	51	136		69
-44		-44	52	165		20

An Aether expressing Fibonacci and Lucas Sequences joined at the hip: a thought experiment

Summary of Fibonacci Carrier and supra-MNS states for the MNS-F(7)^2 circuit

Carrier states and Supra-MNS states for each of the 52 words from the 364 iterations of a full cycle.

- Non-spin corrected carrier states oscillate between "-44" and "125"; and
- Supra-MNS states are generated by multipliers "164" and "135", in essence multipliers of
    - $F(7)^2 - F(5)$ and
    - $F(7)^2 - F(9)$

    with an initial value of "4" i.e., $L(0)^2$.

The progression of carrier states for Fibonacci sequences

As should be noted, the values of the "carrier states" for each MNS-F(n)^2 circuit refer to the same Aether-state, but to the different sides thereof. Thus, as shown:

- For $F(3)^2$ the value is "0" or "4";
- For $F(4)^2$ the value is "4" or "-5";
- For $F(5)^2$ the value is "-5" or "20";
- For $F(6)^2$ the value is "20" or "-44"; and
- For $F(7)^2$ the value is "-44" or "125".

An Aether expressing Fibonacci and Lucas Sequences joined at the hip: a thought experiment

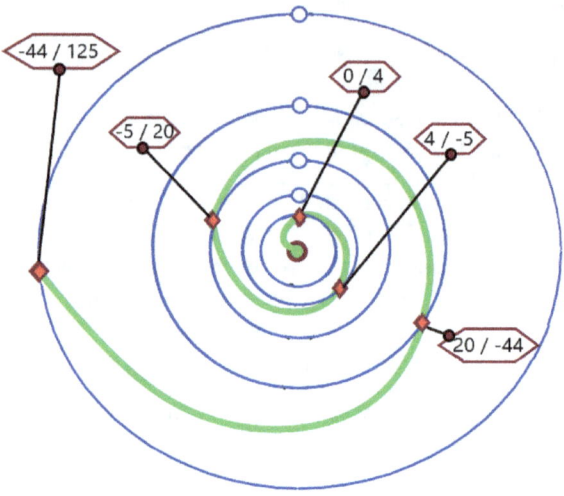

The spiral-like progression of the carrier states through the MNS-F(n)^2 circuits

The progression of the carrier states through the MNS-F(n)^2 circuits follows a simple rule. The Aether-value of the carrier state of the present MNS-F(n)^2 is determined by:

- one of the carrier values of the Aether state of the MNS-F(n-1)^2 circuit and
- one of the carrier values of the Aether state of the MNS-F(n+1)^2 circuit,

An Aether expressing Fibonacci and Lucas Sequences joined at the hip: a thought experiment

wherein the sum of the absolute values of these previous and next carrier states equal $F(n)^2$. Thus if we look at the progression of the values of carrier states, we can see that:

- the lower of the carrier value of the Aether-carrier state of MNS-$F(n)^2$ circuit is equal to the higher carrier value of the Aether-carrier state of the MNS-$F(n-1)^2$ circuit.

Thus, the higher carrier value of the Aether-carrier state of MNS-$F(n)^2$ circuit is merely the counter value of the Aether-carrier state for MNS-$F(n)^2$.

In this respect the carrier values of each MNS-$F(n)^2$ circuit build up upon each other. Starting from "0", and moving through values "4", "-5", "20", "-44", "125", etc. there is a progression of the carrier states which interlink each MNS-$F(n)^2$ circuit from the highest to the lowest. It is proposed therefor that if there is anything which could be described as "zero-point" energy, then this may refer to simply achieving the most efficient way to transfer one form of energy to another form of energy. A path of least resistance.

An Aether expressing Fibonacci and Lucas Sequences joined at the hip: a thought experiment

Celestial bodies subject to the Aether

Up to now I have discussed closed loop number systems and the relationships between closed loop number systems as determined by pushing Fibonacci and Lucas sequences between these systems. Here I will propose relationships between lunar and solar cycles based on the seventh iteration of the Lucas and Fibonacci sequences.

The moon cycle within a month based on an MNS-L(7) Rodin-type force graph.

If we take the data that we have for the moon relevant to this discussion, in particular data relating to the orbital periods and the mass of the moon, then we have [18]:

-   A sidereal orbital period of 27.32 days
-   A synodic orbital period of 29.53 days; and
-   A mass approximately 1/81 that of the Earth.

For the purposes of this paradigm, we will take a cue from the MNS-L(7) circuit, and substituting days for iterations, assume that the moon in a month has 28 Aether iterations.

An Aether expressing Fibonacci and Lucas Sequences joined at the hip: a thought experiment

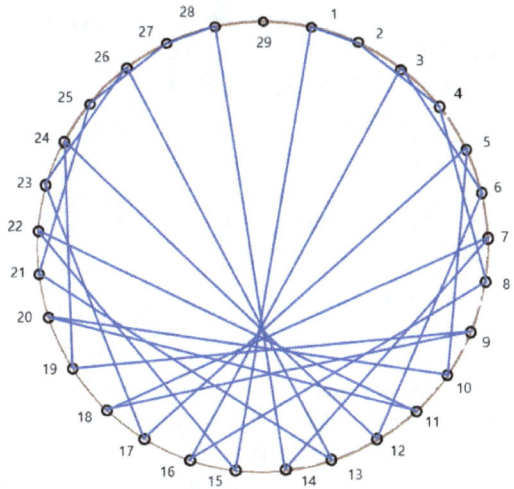

Rodin-type force graph with complementary multipliers "2" and "15"

It is noted that while 28-iterations does not agree with either the measured periods of either the synodic orbital period or the sidereal orbital period, I shall treat the value "28" as being a happy median between the synodic and sidereal orbital periods. Furthermore, in line with the tidal locking of the moon, rather than multipliers of "5" and "6" of least resistance being used, which would be expected for the unbalanced sequence associated therewith of

-    9 _ -4 _ 5 _ **1** _ 6 _ 7,

multipliers of the "2-operator" and "15" will b used.

An Aether expressing Fibonacci and Lucas Sequences joined at the hip: a thought experiment

It is noted in this respect that to obtain the complementary multiplier values of "2" and "15" from the least resistance values "6" and "5", one would be required to:

- Divide "6" by 3 and
- Multiply "5" by 3.

Conversely, it is also noted that in this 28-iteration system derived from the MNS-29 circuit, that if one:

- Divided the mass of the earth by 3 or
- Multiplied the mass of the moon by 3,

then one would obtain a mass ratio of the earth to the moon of 1:27 (rather than 1:81 as is currently estimated). Let's assume the mass of the moon is amended by multiplication by 3. In such a case then, with an earth-moon cyclic system assumed to still have a lunar cycle of 28-iterations, this earth-moon system would have a mass equal to 28 times that of such an amended moon mass. This may be coincidence, but I will let the reader decide.

An Aether expressing Fibonacci and Lucas Sequences joined at the hip: a thought experiment

The Enoch calendar, a year based on an MNS-F(7)^2 circuit.

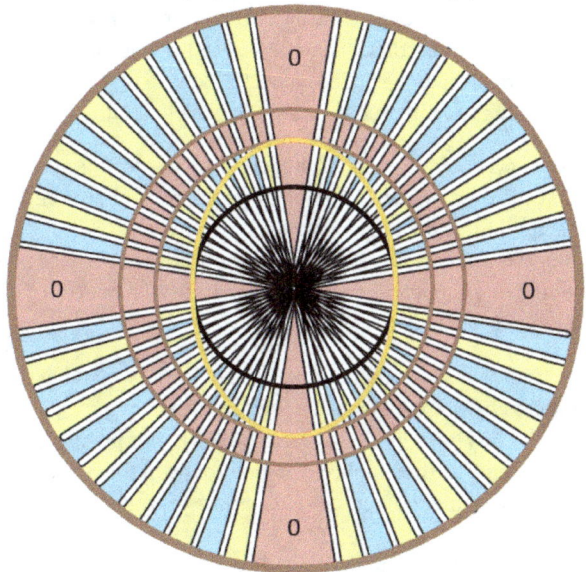

Portrayal of Enoch Calendar in MNS-F(7)^2 [= 169] having 364 iterations but emphasising 52 days of sabbath every seven iterations.

As discussed, the unbalanced MNS-29 circuit has 28 Aether-iterations to complete a full cycle, each Aether-iteration being proposed to be equivalent to one rotation of the earth on its own axis. If we accept the proposition that each moon cycle of 28 iterations is equivalent to a

An Aether expressing Fibonacci and Lucas Sequences joined at the hip: a thought experiment

completed orbit of the moon about the earth in a month, then each moon cycle is equivalent to four weeks of 7 days / iterations of earth. On a numerical sense, it is further noted that the unbalanced value "29" per se is also a Lucas value, the seventh in the Lucas sequence [= L(7)].

In the cycle of the earth orbiting the sun it is further noted that the moon itself, in completing each orbit about the earth, rotates on its own axis approximately 13 [= F(7)] times in a solar year. The question now arises as to what would one expect if one tried to generate a Fibonacci sequence in an MNS-F(7)^2 circuit?

As it turns out, the number of iterations one would have to process before the Fibonacci sequence begins to repeat itself is 364 iterations. This value of 364 is also precisely the same value as the number 364 mentioned in the book of Enoch as being the number of days in a year.

The Book of Enoch is a Hebrew text for which its oldest copies date from 200BC to 300 BC. Several copies have also been found preserved among the dead sea scrolls **[19]**. Taking an excerpt from the Book of Enoch relating to the so-called Enoch calendar:

> "[36.] Into that gate it enters for thirty days, and sets in the west, in the opposite part of heaven.
>
> [37.] At that period the night is contracted in its length a fourth part, that is, one portion, and becomes eleven parts.

An Aether expressing Fibonacci and Lucas Sequences joined at the hip: a thought experiment

38. The day is seven parts.

39. Then the sun returns and enters into the second gate of the east.

40. It *returns by these beginnings thirty days, rising and setting.*

41. At that period the night is contracted in its length. It becomes ten parts, and the day eight parts. Then the sun goes from that second gate and sets in the west; but returns to the east, and rises in the east, *in the third gate, thirty-one days, setting in the west of heaven.*

42. At that period the night becomes shortened. It is nine parts. And the night is equal with the day. *The year is precisely three hundred and sixty-four days.*

43. The lengthening of the day and night, and the contraction of the day and night, are made to differ from each other by the progress of the sun.

44. By means of this progress the day is daily lengthened, and the night greatly shortened."

- Book of Enoch, chapter 71:36-44 **[20]**

In essence the Enoch calendar splits the year into four quarters with each quarter comprising three months ("gates"). Two of the three months of each quarter

193

An Aether expressing Fibonacci and Lucas Sequences joined at the hip: a thought experiment

comprises 30 days and one month comprises 31 days. Thus, each quarter comprises in total 91 days and one year comprises 364 days.

As such the number of days of each month ("gate") differs from the previously mentioned "moon" cycle of 28 days by at least two days, with one month differing by three days. The numerological effect of the addition of these extra days is to in effect synchronise each quarter to cycles of "weeks", where each quarter now is comprised of 13 weeks of seven days precisely. That is, we have the four weeks of each month to give us 12 weeks, and in addition extra days of "2" + "2" + "3" giving an additional 7 days or one additional week for each quarter of the calendar. This may appear quirky at first, until one remembers from the Bible that:

- from genesis it is mentioned that the Lord rested on the seventh day,

    - "Thus the heavens and the earth were finished, and all the host of them.

      [2] And on the seventh day God ended his work which he had made; and he rested on the seventh day from all his work which he had made.

194

An Aether expressing Fibonacci and Lucas Sequences joined at the hip: a thought experiment

> ³ And God blessed the seventh day, and sanctified it: because that in it he had rested from all his work which God created and made."

> Genesis 2:1-3

and that

- from exodus it is codified in the ten commandments that the sabbath day should be held sacred, and that no work should be done on that day.

> - ⁸ Remember the sabbath day, to keep it holy.

> ⁹ Six days shalt thou labour, and do all thy work:

> ¹⁰ But the seventh day is the sabbath of the LORD thy God: in it thou shalt not do any work, thou, nor thy son, nor thy daughter, thy manservant, nor thy maidservant, nor thy cattle, nor thy stranger that is within thy gates:

> ¹¹ For in six days the LORD made heaven and earth, the sea, and all that in them is, and rested the

195

An Aether expressing Fibonacci and Lucas Sequences
joined at the hip: a thought experiment

> seventh day: wherefore
> the LORD blessed the sabbath day,
> and hallowed it."

> \-   Exodus 20:8-11

The question then becomes:

- what is so special about the seventh day / earth
  iteration that the year must be split into four
  quarters of 13 weeks such that each year has an
  integer multiple of weeks?

Other than the heavenly decree received at the foot of
Mt. Sinai, why should every seventh day be so special?

An Aether expressing Fibonacci and Lucas Sequences joined at the hip: a thought experiment

Standard				n	Spin Corrected			
F(x)		F(-x)	F(x) * F(-x)		F(x)		F(-x)	F(x) * F(-x)
	0		0	0		0		0
155		-155	142	82	-14		14	-27
118		118	66	83	-51		-51	66
104		-104	0	84	-65		65	0
53		53	105	85	53		53	-64
157		-157	25	86	-12		12	25
41		41	160	87	41		41	-9
29		-29	4	88	29		-29	4
70		70	168	89	70		70	-1
99		-99	1	90	-70		70	1
0		0	0	91	0		0	0
99		-99	1	92	-70		70	1
99		99	168	93	-70		-70	-1
29		-29	4	94	29		-29	4
128		128	160	95	-41		-41	-9
157		-157	25	96	-12		12	25
116		116	105	97	-53		-53	-64
104		-104	0	98	-65		65	0
51		51	66	99	51		51	66
155		-155	142	100	-14		14	-27

Portion from the MNS-169 table indicating the 82nd to 100th iterations.

Let's look at an excerpt from a table listing the iterations of the 364-day cycle of the MNS-F(7)^2 circuit. In this case it will be part of the sequence about iteration 91 where generated values are transitioning from the first quarter to the second quarter of the year. Bear in mind that each Aether-value of 13, and integer multiples thereof, are now the equivalent of root-"0" as a form of pseudo-dualling numbers. At the 84th and 98th iterations we can see, whether standard or spin-corrected, the products of F(x) * F(-x) give rise to an Aether-value of "0". On the other hand,

An Aether expressing Fibonacci and Lucas Sequences joined at the hip: a thought experiment

at the 91st iteration, the values of F(x) and F(-x) are already equal to "0", and so the product of F(x) * F(-x) is 0^2. The 91st iteration is also the endpoint and start the point of successive palindromes of 92-iteration length. This relationship between the 91st iteration and the product of F(x) and F(-x) is true also for the 182nd iteration, the 273rd iteration and the 364th iteration. That is for each of the 0th, 91st, 182nd, 273rd, and 364th iterations:

- the Aether-values of the complementary time strands F(x) and F(-x) are each equal to "0"; and
- each serve as the endpoint and start the point of successive palindromes of 92-iteration length.

While I cannot unambiguously conclude that keeping a calendar which synchronises with cycles of shabbat is based on the Fibonacci sequence being driven through an MNS-F(7)^2 circuit, its hard to argue that something like the MNS-F(7)^2 circuit was used as a template for the Enoch calendar. Further, as the earliest records for the Enoch calendar go back at least 2100 years, it raises the question as to whether the teaching of this book in relation to Fibonacci and Lucas sequences is an invention or re-discovery.

The Enoch Calendar, a calendar of Shabbats.
The question now arises as to why anyone would find it important to keep a calendar of shabbats? ... or why the Lord himself would put such emphasis on keeping the shabbat? Indeed, if we read for example a passage from the book of Jeremiah :

198

An Aether expressing Fibonacci and Lucas Sequences joined at the hip: a thought experiment

"21 Thus saith the Lord; Take heed to yourselves, and *bear no burden on the sabbath day*, nor bring it in by the gates of Jerusalem;

22 *Neither carry forth a burden out of your houses on the sabbath day, neither do ye any work, but hallow ye the sabbath day, as I commanded your fathers.*

23 But they obeyed not, neither inclined their ear, but made their neck stiff, that they might not hear, nor receive instruction.

24 And it shall come to pass, if ye diligently hearken unto me, saith the Lord, to bring in no burden through the gates of this city on the sabbath day, *but hallow the sabbath day, to do no work therein*;

25 Then shall there enter into the gates of this city kings and princes sitting upon the throne of David, riding in chariots and on horses, they, and their princes, the men of Judah, and the inhabitants of Jerusalem: and this city shall remain for ever.

26 And they shall come from the cities of Judah, and from the places about Jerusalem, and from the land of Benjamin, and from the plain, and from the mountains, and from the south, bringing burnt offerings, and sacrifices, and meat

An Aether expressing Fibonacci and Lucas Sequences joined at the hip: a thought experiment

offerings, and incense, and bringing sacrifices of praise, unto the house of the Lord.

27 But if ye will not hearken unto me to hallow the sabbath day, and not to bear a burden, even entering in at the gates of Jerusalem on the sabbath day; *then will I kindle a fire in the gates thereof, and it shall devour the palaces of Jerusalem, and it shall not be quenched."*

- Jeremiah 17:22-27

an attitude of the kingdom of Judah to ignore the sabbath, and thereby pollute it, was tantamount to *casus belli* for the Lord to bring punishment to his people. But let's look at the table for the MNS-F$(7)^2$ circuit and observe the products of time-strands F(x) and F(-x) every seven iterations. It is the word punctuation point which defines the "word" of the Aether carrier-state for MNS-F$(7)^2$, and it is a value of "0". I propose that this value of "0" in the MNS-F$(7)^2$ circuit, resonates with the "0" at the beginning of time in which the Lord spoke the first words to bring forth all of creation. It is those days of sabbath where people through doing nothing, come closest to being partakers in the story of creation. In response to the Lords question to Job in the book of Job of

- "*4 Where wast thou when I laid the foundations of the earth? declare, if thou hast understanding.
- *5 Who hath laid the measures thereof, if thou knowest? or who hath stretched the line upon it?

200

An Aether expressing Fibonacci and Lucas Sequences joined at the hip: a thought experiment

- [6] Whereupon are the foundations thereof fastened? or who laid the corner stone thereof;
- [7] When the morning stars sang together, and all the sons of God shouted for joy?"
  - Job 38:4-7

can keepers of the sabbath respond:

- We were with You Lord?

In short for purposes spirituality and consciousness, the sabbath is the day / iteration within the calendar with which we can seek to attune or realign ourselves with God. Where our minds are not subject to pleasure seeking as recreation but finding pleasure in re-creating our minds and souls in resonance with the original words of creation. Nothing spoken by the Lord is in vain.

Admittedly the number of days within the Enoch calendar is shorter than the number of days within either the synodic year (365.25) or the sideral year (366.25) [21] by values of 1.25 and 2.25 days. However, as:

- this difference in length of year, in terms of sidereal days at least, approximates "2"; and
- the "2-operator" of L(0) emerges frequently in the tables detailing Fibonacci harmonic systems,

I am open minded as to whether this difference of "2" days / iterations between

An Aether expressing Fibonacci and Lucas Sequences joined at the hip: a thought experiment

- the Enoch year and
- the year of sidereal days

may not serve so much as a bug but rather a feature.

Lastly, it should further be noted that while a year on earth is ultimately measured as the time for the earth to complete an orbit about the sun, the sun itself is not stationary. Indeed, depending upon its latitude, the sun spins on its own axis from every 24.47 days (at its equator) to 38 days (at its poles) **[22]**. There are therefor, presumably upper, and lower latitudes of the sun for which the orbital cycle of the moon is synchronised (or is anti-synchronised with depending on direction of rotation of the sun) with the rotation of some latitudes of the sun. Thus, for the sun, the cycles in which the earth orbits of the sun (fur us a year) are just about thirteen "days" (or approximately the number of moon cycles in a terrestrial year).

An Aether expressing Fibonacci and Lucas Sequences joined at the hip: a thought experiment

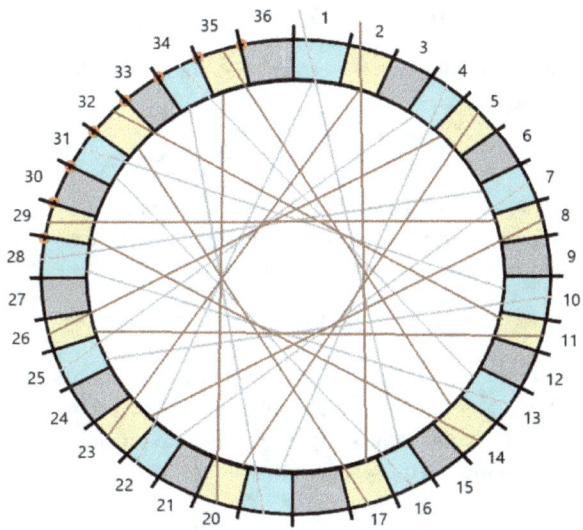

Stylistic Rodin coil with two wires wound once in a circular configuration.

Above is an example of a Rodin coil **[16, 17]** in which first (grey) and second (brown) copper wires are wrapped at 12 points about a circle. As of the time of writing, I have not created a Roding coil for myself but apparently, many who have created Rodin coils tend to wrap each wire eighteen times about the circuit.

203

An Aether expressing Fibonacci and Lucas Sequences joined at the hip: a thought experiment

Building of a Rodin coil

Thus, the grey line wire would follow the sequence:

1_16_31_10 _25_4_19_34_13_28_7_22

eighteen times, and the brown line wire would follow the sequence:

2 _17_32_11_26_5_20_35_14_29_8_ 23

eighteen times. I am unaware as to whether the wires are wound separately, or simultaneously to achieve an interlaced structure. The placements:

3_6_9_12_15_18_21_21_24_27_30_33_36

are left vacant. From the YouTube (RTM) videos I have viewed, the vacancies are for allowing the circuit to breath. From my point of view these gaps serve to provide punctuation of the words of the circuit.

Possible words emerging from the Rodin coil structure.

As far as a I can tell from the various YouTube (RTM) videos where Mr. Marko Rodin is either teaching or giving interviews, great emphasis is laid on the number sequence:

- 1 _ 2 _ 4 _ 8 _ 7 _ 5

which repeats every six iterations in the decimal system. Further iterations of the sequence are reduced to "1", "2", "4", "8", "7" and "5" through addition of digits of the decimal numbers obtained through multiplication by "2" to

204

An Aether expressing Fibonacci and Lucas Sequences joined at the hip: a thought experiment

reduce the values to a single digit. In effect a pseudo modulo-9 number system. This sequence is attributed to the operator of "2" (also a multiplier $F(3)$ in the Cassini Identity for $F(4)^2$) acting as multiplier starting from an initial value of "1". A reverse of the sequence is also achieved using the complementary multiplier of $F(5)$ of the Cassini Identity for $F(4)^2$. As it happens, the sequence of:

- 1 _ 2 _ 4 _ 8 _ 7 _ 5

also emerges for each of the individual time-strands $F(x)$ and $F(-x)$ in an MNS-$F(4)^2$ circuit [MNS-9 circuit], or an MNS-$L(2)^2$ circuit, i.e., when using $F(4)$ [= 3] or $L(2)$ [= 3] as punctuation. In the latter case of the MNS-$L(2)^2$ circuit, there is a shift of 30-degrees relative to the application of punctuation in the MNS-$F(4)^2$ circuit.

Carrier States				Supra-MNS states		
Original Sum		Corrected Sum	Word Index n	Sum (x+)		Sum (x-)
	0				0	
4		4	1	4		2
-5		4	2	8		1
4		4	3	7		5
4		4	4	5		7
-5		4	5	1		8
4		4	6	2		4

Carrier states and Supra-MNS states for the Fibonacci Sequence based on MNS-$F(4)^2$ / MNS-$L(2)^2$ circuit.

An Aether expressing Fibonacci and Lucas Sequences joined at the hip: a thought experiment

Carrier states				Supra-MNS states		
Original Sum		Corrected Sum	Word Index n	Sum (r+)		Sum (r-)
	4				4	
2		2	1	2		2
2		2	2	4		1
2		2	3	8		5
2		2	4	7		7
2		2	5	5		8
2		2	6	1		4
2		2	7	2		2

Carrier states and Supra-MNS states for the Lucas Sequence based on MNS-F(4)^2 / MNS-L(2)^2 circuit.

If we now assume that one wire comprises electrons within the copper lattice of the wire for which values of the sequence represent states of electron pairs within that conductor, and that the other wire also comprises electrons within the copper lattice of that wire for which values of the sequence represent states of electron pairs within that conductor, the question arises, what can be achieved by winding these two wires in loops such that they crossover one another in a Rodin coil fashion?

Lattice structure of copper leading to the emergence of MNS-9 electron circuits

It is the supposition of the Author that copper metal, being Face Centred Cubic (FCC), involves each copper atom sharing twelve of its electrons with respective ones of twelve neighbouring atoms. Thus, each copper atom at its connection to its neighbouring atom would share a respective pair of electrons. The consequence thereof would be that each copper atom, as part of the lattice, would have five electrons more than those required for achieving the full shell structure of the noble gas krypton. These five surplus electrons however could not be expelled from the lattice as the charge neutrality needed to be maintained.

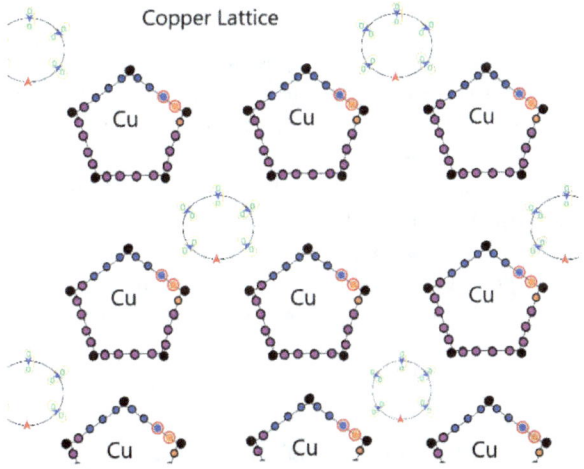

Copper lattice illustrating a copper shell based on F(5)^2 comprising F(7) electron pairs and expelled electrons forming a structure of F(5) electron pairs.

It is submitted therefor that these excess electrons of the copper atoms would combine with the surplus electrons

An Aether expressing Fibonacci and Lucas Sequences joined at the hip: a thought experiment

from a neighbouring atom, and thereby form a neon type electron circuit structure. Furthermore, it is submitted that this neon-type structure of five pairs of electrons is subject to rules of both the MNS-F(4)^2 and MNS-L(2)^2 circuits.It is the further premise that each electron pair of the neon-like structure has associated therewith, at any one time, one of six Aether states of the 1-2-4-8-7-5 sequence. Conversely the neon-type electron circuit may be defined by the single Aether state of the possible six Aether states which is not occupied by an electron pair.

It should also be noted that copper per se can also be demonstrably shown to have magnetic properties, if rather weak in comparison to hose displayed by iron. It is submitted that these properties of the lattice may be a result of the author-postulated F(7) * F(3) pairs of electrons which result out of the FCC structure of copper. It is submitted also that these low-level magnetic properties may not only be properties of copper, but that they may also properties of the other main conductors of silver and gold.

Possible outcomes of the complementary wires

As far as can be discerned by the Author, the Rodin coil structure in the centre of the circle is trying to recreate the neon-type electron structures within the conductor itself. Pushing the 1-2-4-8-7-5 sequence within a single wire within the structure however is not considered to produce any further effect beyond that which is already exhibited by the conductor if it were not wound in the coil. I have not

An Aether expressing Fibonacci and Lucas Sequences joined at the hip: a thought experiment

carried out any experiments for this, so I cannot determine whether my assumption in this respect is true.

On the other hand, introducing a second coil allows an opportunity for something to be produced through interactions between states of the first wire and the second wire. For example, if the phase of between the wires could be manipulated so that the values could interact with one another such that:

A   B

-   1 * 4 = 4   1 * 2 = 2
-   2 * 2 = 4   2 * 1 = 2
-   4 * 1 = 4   4 * 5 = 2 [=
    20 mod 9]
-   8 * 5 = 4 [= 40 mod 9]   8 * 7 = 2 [=
    56 mod 9]
-   7 * 7 = 4 [= 49 mod 9]   7 * 8 = 2 [=
    56 mod 9]
-   5 * 8 = 4 [= 40 mod 9]   5 * 4 = 2 [=
    20 mod 9]

It is submitted that in the case of A), if we supply the second wire with a sequence opposite to that of the first wire and shifted by two steps, then we will create a "carrier"-state of "4" which is the Lucas complement of $F(0)^2$ in the Fibonacci sequence.

In the case of B), if we supply the second wire with a sequence opposite to that of the first wire and shifted by

An Aether expressing Fibonacci and Lucas Sequences joined at the hip: a thought experiment

one step, then we will create a "carrier"-state of "2" which is the Lucas complement of F(0) in the Fibonacci sequence.

In the case we supply the second wire with a sequence opposite to that of the first wire and shifted by a zero step, then we will create a "carrier"-state of "1".

Thus, with whatever reverse sequence one applies to the second wire relative to the first wire, one achieves a "carrier"-state rather than any sequence of Supra-MNS states. The question then arrives in what respect any of these "carrier"-states enables further interaction for the drawing of current or the generation of magnetic fields.

The reader is reminded that all these ideas of this book regarding the Rodin coil are speculative. I am writing here not to procure answers, but to provide pertinent questions.

An Aether expressing Fibonacci and Lucas Sequences joined at the hip: a thought experiment

Applying the Aether-based Fibonacci equation to the Aether-based resonant circuit

Now let's go back to **EQN(5)**, if we factorize **EQN(6)**

$$F(n) = F\left(\frac{(2n + 3 + i^{2n})}{4}\right)^2 - i^{2n}F\left(\frac{(2n - 3 - i^{2n})}{4}\right)^2$$

we get the following factors:

$$F(n) =$$

$$F\left(\frac{2n + 3 + i^{2n}}{4}\right) + i^n F\left(\frac{2n - 3 - i^{2n}}{4}\right) \times$$

$$F\left(\frac{2n + 3 + i^{2n}}{4}\right) - i^n F\left(\frac{2n - 3 - i^{2n}}{4}\right)$$

If we now treat the above upper / left and lower / right components of the factorised **EQN 5** as corresponding impedances for a pseudo-inductor (ZL) and a pseudo-capacitor (ZC), we get:

$$- \quad ZL = \frac{F\left(\frac{2n + 3 + i^{2n}}{4}\right) + i^n F\left(\frac{2n - 3 - i^{2n}}{4}\right)}{}$$

and

An Aether expressing Fibonacci and Lucas Sequences joined at the hip: a thought experiment

$$- \quad ZC = F\left(\frac{2n + 3 + i^{2n}}{4}\right) - i^n F\left(\frac{2n - 3 - i^{2n}}{4}\right)$$

whereby

$$- \quad RL = RC = F\left(\frac{2n + 3 + i^{2n}}{4}\right) = RAe$$

and

$$- \quad XL = - XC = i^n F\left(\frac{2n - 3 - i^{2n}}{4}\right) = XAe$$

where RAe and XAe are the respective resistance and reactance components of the elements of this hypothetical tank circuit representing properties of the Aether.

Thus, the Aether impedance ZAe becomes:

$$ZAe = \frac{(RAe + XAe) \times (RAe - XAe)}{2RAe}$$

This however is where the analogy for the Aether ends as, for actual capacitors and inductors, "n" only has a value of 1.

An Aether expressing Fibonacci and Lucas Sequences joined at the hip: a thought experiment

For example, substituting back the Aether values to the Resonant Circuit analogy of **EQN(6)** gives us the value:

ZAe(1)

$$= \left(\frac{1}{2 \times F(1)}\right) \times \left(F(1)^2 + F(0)^2\right)$$

$$= \left(\frac{1}{2}\right)$$

From the above aspect however for "n" = 1, it can be seen the value of ½ appears to modulate the "impedance" value of the Aether.

An Aether expressing Fibonacci and Lucas Sequences joined at the hip: a thought experiment

A graphical presentation of a sequence of Fibonacci numbers in the form of a progression of arrayed squares

Green = **Negative**
Blue = **Positive**

F(-4) = -3   F(4) = 3

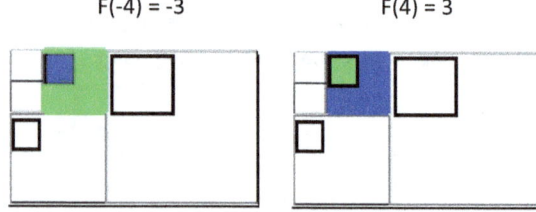

An Aether expressing Fibonacci and Lucas Sequences joined at the hip: a thought experiment

$F(-9)$	$= F(-4)^2 + F(-5)^2$	$= 9 + 25$	$= 34$
$F(-8)$	$= F(-3)^2 - F(-5)^2$	$= 4 - 25$	$= -21$
$F(-7)$	$= F(-3)^2 + F(-4)^2$	$= 4 + 9$	$= 13$
$F(-6)$	$= F(-2)^2 - F(-4)^2$	$= 1 - 9$	$= -8$
$F(-5)$	$= F(-2)^2 + F(-3)^2$	$= 1 + 4$	$= 5$
$F(-4)$	$= F(-1)^2 - F(-3)^2$	$= 1 - 4$	$= -3$
$F(-3)$	$= F(-1)^2 + F(-2)^2$	$= 1 + 1$	$= 2$
$F(-2)$	$= F(0)^2 - F(-2)^2$	$= 0 - 1$	$= -1$
$F(-1)$	$= F(0)^2 + F(-1)^2$	$= 0 + 1$	$= 1$
$F(0)$	$= F(1)^2 - F(-1)^2$	$= 1 - 1$	$= 0$
$F(1)$	$= F(1)^2 + F(0)^2$	$= 1 + 0$	$= 1$
$F(2)$	$= F(2)^2 - F(0)^2$	$= 1 - 0$	$= 1$
$F(3)$	$= F(2)^2 + F(1)^2$	$= 1 + 1$	$= 2$
$F(4)$	$= F(3)^2 - F(1)^2$	$= 4 - 1$	$= 3$
$F(5)$	$= F(3)^2 + F(2)^2$	$= 4 + 1$	$= 5$
$F(6)$	$= F(4)^2 - F(2)^2$	$= 9 - 1$	$= 8$
$F(7)$	$= F(4)^2 + F(3)^2$	$= 9 + 4$	$= 13$
$F(8)$	$= F(5)^2 - F(3)^2$	$= 25 - 4$	$= 21$
$F(9)$	$= F(5)^2 + F(4)^2$	$= 25 + 9$	$= 34$

An Aether expressing Fibonacci and Lucas Sequences joined at the hip: a thought experiment

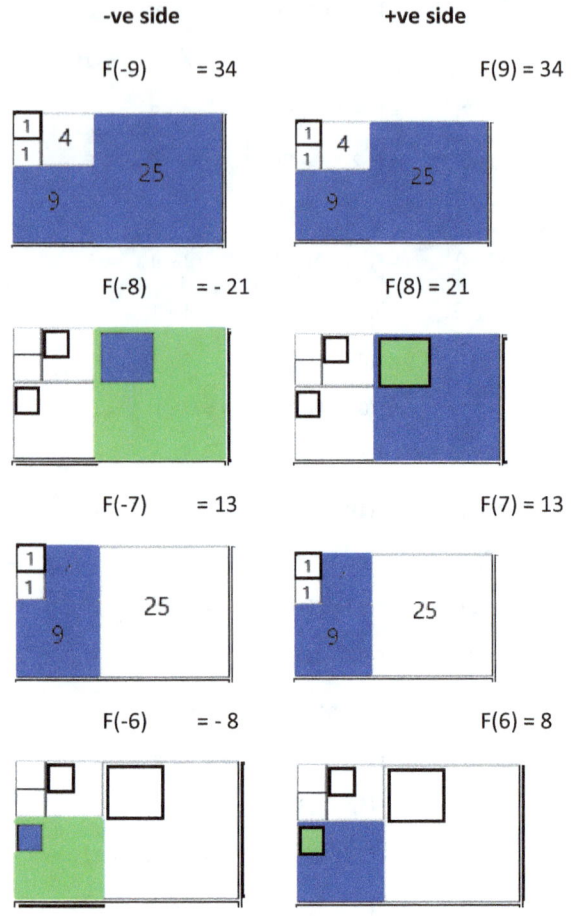

An Aether expressing Fibonacci and Lucas Sequences joined at the hip: a thought experiment

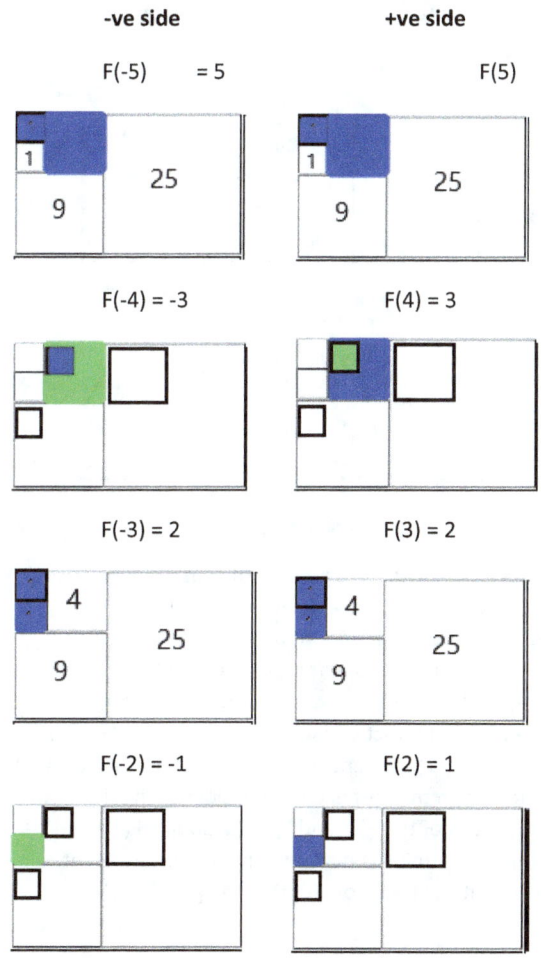

An Aether expressing Fibonacci and Lucas Sequences joined at the hip: a thought experiment

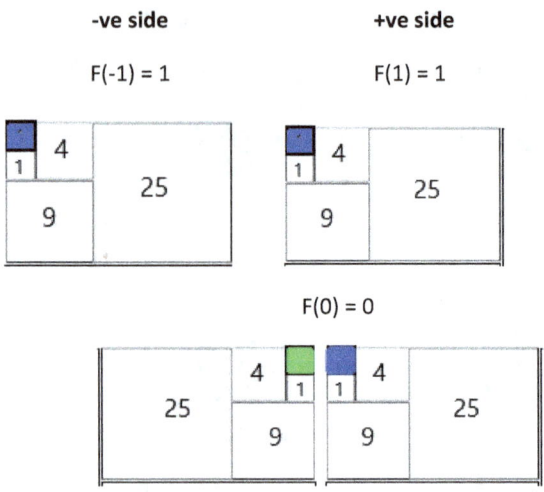

As can be seen by the progression of squares in both left/(-ve)negative and right/(+ve)positive sides about the boundary, calculations within the shadow/reactive and observer/real spaces appear independent of one another. It is only at the boundary nexus of F(0) where shadow and observer sides interact with one another. Indeed it is only the F(1) and F(-1) elements of each side which interact with one another. In this respect, it is considered that it is also the properties of the F(1) and F(-1) (including F(1)^2 and F(-1)^2) which determine the characteristics of the Aether vortex about the nexus of F(0) (including F(0) x L(0)).

An Aether expressing Fibonacci and Lucas Sequences joined at the hip: a thought experiment

Epilogue 3:

An alternate means of determining odd values of the Fibonacci sequence

We have already seen computational means by which values of F(n) may be calculated where "n" is even e.g.,

- $F(n) = F(n/2) * L(n/2)$

or even

- $F(n) = F(n/2 +1)^2 - F(n/2 - 1)^2$

If one also follows notes the sequence of arrayed squares illustrating the progression of the Fibonacci sequences from an origin, one also notes that for each *odd* value of "n", the corresponding element value F(n) may be calculated by a difference of products of a set of a sequential set of four previous Fibonacci values. That is for a Fibonacci value of F(2X − 3), this is equivalent to:

$$F(2X - 3)$$
$$= (F(X) * F(X - 1)) - (F(X - 2) * F(X - 3))$$

where "X" is an even value.

To determine a value of an odd value of "n" of the expression F(n), an expression must be used to determine a value of "X" based on the value of "n" which we wish to determine. That is we start off with

$$n = 2X - 3$$

An Aether expressing Fibonacci and Lucas Sequences joined at the hip: a thought experiment

whereby. To determine "X" we rewrite the expression to

$$X = (n + 3)/2$$

For example, if we wish to determine F(n) e.g. where "n" = 9, F(1), then

$$X = (11 + 3)/2 = "7",$$

whereby calculating the value of F(11) becomes:

$$F(11) \quad = (F(7) * F(6)) - (F(5) * F(4))$$

$$= (13 * 8) - (5 * 3)$$

$$= 104 - 15$$

$$= 89$$

Conversely, in determining the value of F(-11) the expression still holds with

$$"X" = (n + 3)/2 = (-11 + 3)/2 = "-4",$$

Whereby calculating the value of F(-11) becomes:

$$F(-11) \quad = (F(-4) * F(-5)) - (F(-6) * F(-7))$$

$$= ((-3) * (-5)) - ((-8) * (13))$$

$$= -15 - (-104)$$

$$= 89$$

At low values of "n" it may be computationally efficient to simply add the previous numbers in the Fibonacci sequence

An Aether expressing Fibonacci and Lucas Sequences joined at the hip: a thought experiment

to obtain the value of F(n+1), i.e., according to F(n+1) = F(n) + F(n-1). But at higher numbers of "n" for e.g., n = 103, it may be useful to base a calculation for F(103) based on

- already calculated values of F(53) and F(51)

from which

- values for n = 52 [F(53) – F(51)] and n = 50 [F(52) – F(51)] may also be calculated.

One could also think that with the huge numbers which F(53 and F(51) may encompass, it may still be more computationally efficient to progress.

- F(n+1) = F(n) + F(n-1).

An Aether expressing Fibonacci and Lucas Sequences joined at the hip: a thought experiment

Epilogue 4:

Spin corrected and non-spin corrected values for the Fibonacci sequence and Lucas sequence.

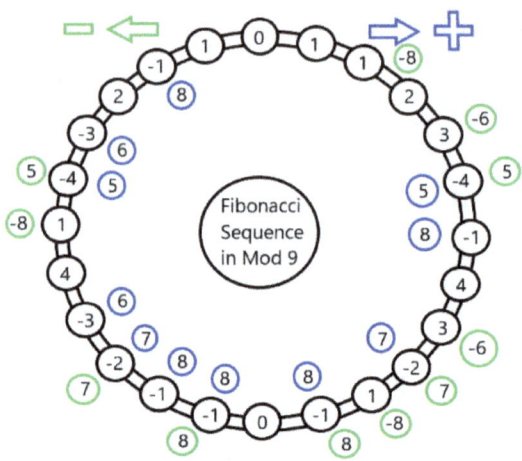

Spin Corrected Aether States of the Fibonacci sequence in MNS-9 showing changes for positive and negative "n"-values

As can be seen above, for "spin correction" what is required is that:

- If a positive value is greater than 9/2, then subtract 9 to provide a corresponding negative number; and

An Aether expressing Fibonacci and Lucas Sequences joined at the hip: a thought experiment

- If a negative value is less than -9/2, then add 9 to provide a corresponding positive number.

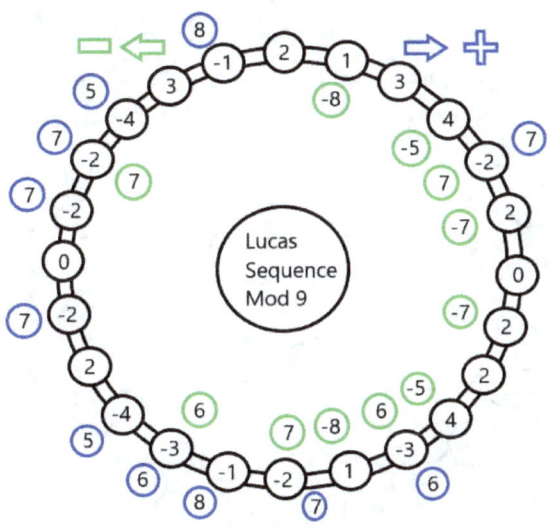

Spin Corrected Aether States of the Lucas sequence in MNS-9 showing changes for positive and negative "n"-values

An Aether expressing Fibonacci and Lucas Sequences joined at the hip: a thought experiment

Dividing values of the MNS-9 based Lucas sequence by the Aether operator "2".

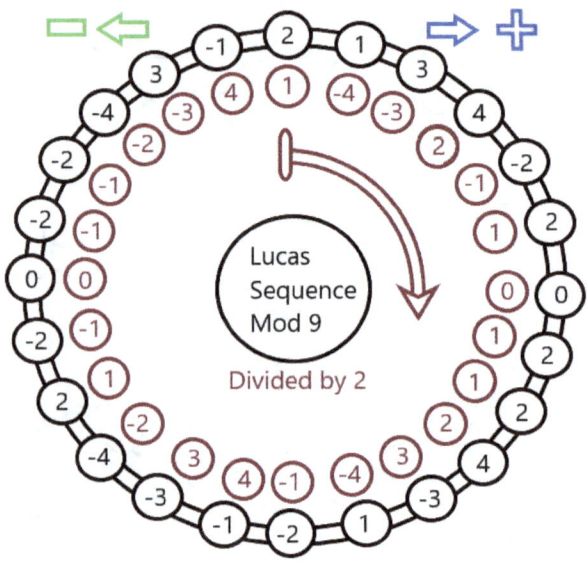

Spin corrected Lucas Sequence in MNS-F(4)² and Spin corrected Lucas sequence in MNS-F(4)² divided by F(3) [= 2] or multiplied by F(5) [= 5]

An Aether expressing Fibonacci and Lucas Sequences joined at the hip: a thought experiment

When dividing

- even numbers in the modulo-9 [i.e., $F(4)^2$] circuit by a value of "2" [i.e., $F(3)$], the result is simply that even value divided by "2". That is
    - division of "2" or "4" by "2" gives respective values "1" and "2";

and when dividing

- odd numbers in the modulo-9 [i.e., $F(4)^2$] circuit by a value of "2" [i.e., $F(3)$], the result achieved requires adding "9" to that value [i.e. equivalent to adding "0" to the value] and then dividing the resultant even number by "2", and then followed by spin correction if required. That is division of "1" or "3" by "2" requires:
    - firstly, the adding of "9" to achieve respective values of "10" and "12";
    - secondly, the dividing of values of "10" and "12" by "2" to achieve values of "5" and "6"; and
    - spin-correcting the values "5" and "6" to result in values "-4" and "-3".

When achieving "division" by the multiplication of numbers in the modulo-9 [i.e., $F(4)^2$] circuit by a value of "5" [i.e., the $F(5)$ multiplier from the Cassini Identity for $F(4)$], the values achieved result simply from:

- multiplication of the numbers of the values by "5"

An Aether expressing Fibonacci and Lucas Sequences joined at the hip: a thought experiment

-   reducing the achieved number to its
    corresponding MNS value;

and

-   if necessary, subjecting the resulting MNS value
    to spin correction. Thus
    -   1 x 5 = 5 = -4   [spin
        corrected]
    -   2 x 5 = 10 = 1
    -   3 x 5 = 15 = 6 = -3   [spin corrected]
    -   4 x 5 = 20 = 2,

whereby initial input values to be divided of "1", "2", "3"
and "4", when subject to multiplication by "5", give rise to
divided values of "4", "1", "-3" and "2" respectively.

An Aether expressing Fibonacci and Lucas Sequences joined at the hip: a thought experiment

Epilogue 6:

Rodin-type diagrams with MNS-F(5)^2 Aether circuit using Cassini identity multipliers "3" and "17" as well as "8" and "22"

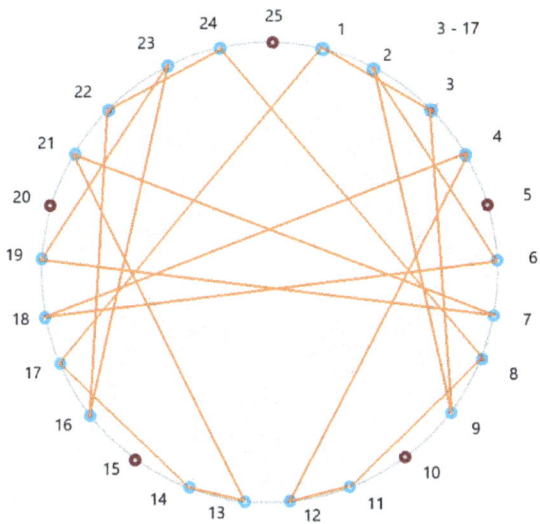

Rodin-type diagram for modulo-25 with complementary multipliers and divisors of

- **"3" [or F(4)] with "17" [25 + (− 8)[or F(-6) = 17].**

An Aether expressing Fibonacci and Lucas Sequences joined at the hip: a thought experiment

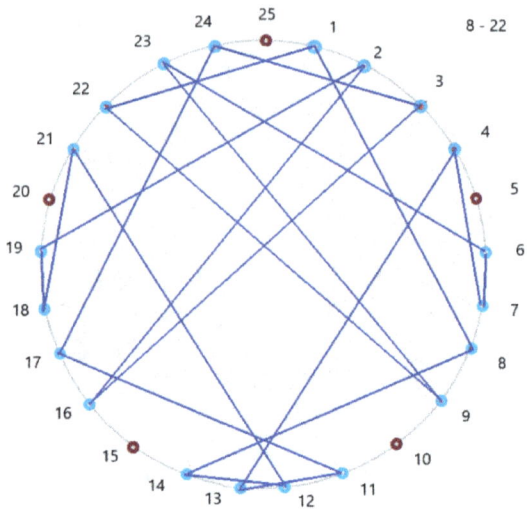

Rodin-type diagram for modulo-25 with complementary multipliers and divisors of

- **"8" [or F(6)] with "22" [25 + (− 3) [or F(-4)] = 22].**

An Aether expressing Fibonacci and Lucas Sequences
joined at the hip: a thought experiment

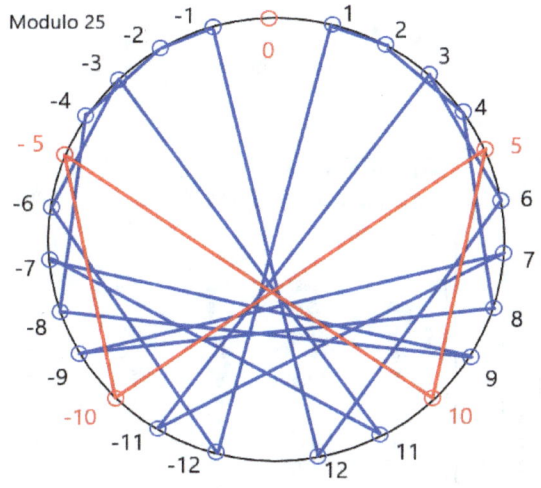

**Rodin-type diagram for Modulo-25 [F(5)²] with
multipliers of "2" [F(3)] and "13" [F(7)] as per
Catalan Identity where "r" = 2**

An Aether expressing Fibonacci and Lucas Sequences joined at the hip: a thought experiment

Epilogue 7

Examples of normalised tables for component tables of unbalanced sequences using "balanced" component table

A case of an "unbalanced" algebraic sequence with

- $U(0) = 1$ and
- $U(1) = 6$ wherein the Cassini Identity = -29

n	F(n)	F(n) * (-1) * F(-n)	F(n-1) * F(-(n+1))	F(n-2) * (-1) * F(-(n+2))	F(n-3) * F(-(n+3))	F(n-4) * (-1) * F(-(n+4))	F(n-5) * F(-(n+5))
-8	-92	12788					
-7	57	-4902	-4865				
-6	-35	1855	1892	1840			
-5	22	-726	-689	-741	-644		
-4	-13	260	297	245	342	-139	
-3	9	-117	-80	-132	-35	-516	368
-2	-4	28	65	13	110	-371	513
-1	5	-30	7	-45	52	-429	455
0	1	-1	36	-16	81	-400	484
1	6	-30	7	-45	52	-429	455
2	7	28	65	13	110	-371	513
3	13	-117	-80	-132	-35	-516	368
4	20	260	297	245	342	-139	
5	33	-726	-689	-741	-644		
6	53	1855	1892	1840			
7	86	-4902	-4865				
8	139	12788					

Component Table

An Aether expressing Fibonacci and Lucas Sequences joined at the hip: a thought experiment

For "r" = 0

n	F(n)	F(n) * (-1) * F(-n)	F(n-1) * F(-(n+1))	F(n-2) * (-1) * F(-(n+2))	F(n-3) * F(-(n+3))	F(n-4) * (-1) * F(-(n+4))	F(n-5) * F(-(n+5))
-3	9	0	37	-15	82	-399	485
-2	-4	0	37	-15	82	-399	485
-1	5	0	37	-15	82	-399	485
0	1	0	37	-15	82	-399	485
1	6	0	37	-15	82	-399	485
2	7	0	37	-15	82	-399	485
3	13	0	37	-15	82	-399	485

For "r" = 1

n	F(n)	F(n) * (-1) * F(-n)	F(n-1) * F(-(n+1))	F(n-2) * (-1) * F(-(n+2))	F(n-3) * F(-(n+3))	F(n-4) * (-1) * F(-(n+4))	F(n-5) * F(-(n+5))
-3	9	-37	0	-52	45	-436	448
-2	-4	-37	0	-52	45	-436	448
-1	5	-37	0	-52	45	-436	448
0	1	-37	0	-52	45	-436	448
1	6	-37	0	-52	45	-436	448
2	7	-37	0	-52	45	-436	448
3	13	-37	0	-52	45	-436	448

For "r" = 2

n	F(n)	F(n) * (-1) * F(-n)	F(n-1) * F(-(n+1))	F(n-2) * (-1) * F(-(n+2))	F(n-3) * F(-(n+3))	F(n-4) * (-1) * F(-(n+4))	F(n-5) * F(-(n+5))
-3	9	15	52	0	97	-384	500
-2	-4	15	52	0	97	-384	500
-1	5	15	52	0	97	-384	500
0	1	15	52	0	97	-384	500
1	6	15	52	0	97	-384	500
2	7	15	52	0	97	-384	500
3	13	15	52	0	97	-384	500

231

An Aether expressing Fibonacci and Lucas Sequences joined at the hip: a thought experiment

For "r" = 3

n	F(n)	F(n) * (-1) * F(-n)	F(n-1) * F(-(n+1))	F(n-2) * (-1) * F(-(n+2))	F(n-3) * F(-(n+3))	F(n-4) * (-1) * F(-(n+4))	F(n-5) * F(-(n+5))
-3	9	-82	-45	-97	0	-481	403
-2	-4	-82	-45	-97	0	-481	403
-1	5	-82	-45	-97	0	-481	403
0	1	-82	-45	-97	0	-481	403
1	6	-82	-45	-97	0	-481	403
2	7	-82	-45	-97	0	-481	403
3	13	-82	-45	-97	0	-481	403

For "r" = 4

n	F(n)	F(n) * (-1) * F(-n)	F(n-1) * F(-(n+1))	F(n-2) * (-1) * F(-(n+2))	F(n-3) * F(-(n+3))	F(n-4) * (-1) * F(-(n+4))	F(n-5) * F(-(n+5))
-3	9	399	436	384	481	0	884
-2	-4	399	436	384	481	0	884
-1	5	399	436	384	481	0	884
0	1	399	436	384	481	0	884
1	6	399	436	384	481	0	884
2	7	399	436	384	481	0	884
3	13	399	436	384	481	0	884

For "r" = 5

n	F(n)	F(n) * (-1) * F(-n)	F(n-1) * F(-(n+1))	F(n-2) * (-1) * F(-(n+2))	F(n-3) * F(-(n+3))	F(n-4) * (-1) * F(-(n+4))	F(n-5) * F(-(n+5))
-3	9	-485	-448	-500	-403	-884	0
-2	-4	-485	-448	-500	-403	-884	0
-1	5	-485	-448	-500	-403	-884	0
0	1	-485	-448	-500	-403	-884	0
1	6	-485	-448	-500	-403	-884	0
2	7	-485	-448	-500	-403	-884	0
3	13	-485	-448	-500	-403	-884	0

An Aether expressing Fibonacci and Lucas Sequences joined at the hip: a thought experiment

A case of an "unbalanced" algebraic sequence with

- U(0) = 1 and
- U(1) = 4 wherein the Cassini Identity = -11

n	F(n)	F(n) * (-1) * F(-n)	F(n-1) * F(-(n+1))	F(n-2) * (-1) * F(-(n+2))	F(n-3) * F(-(n+3))	F(n-4) * (-1) * F(-(n+4))	F(n-5) * F(-(n+5))
-8	-50	4850					
-7	31	-1860	-1843				
-6	-19	703	720	700			
-5	12	-276	-259	-279	-250		
-4	-7	98	115	95	124	-97	
-3	5	-45	-28	-48	-19	-240	100
-2	-2	10	27	7	36	-185	155
-1	3	-12	5	-15	14	-207	133
0	1	-1	16	-4	25	-196	144
1	4	-12	5	-15	14	-207	133
2	5	10	27	7	36	-185	155
3	9	-45	-28	-48	-19	-240	100
4	14	98	115	95	124	-97	
5	23	-276	-259	-279	-250		
6	37	703	720	700			
7	60	-1860	-1843				
8	97	4850					

Component Table

233

An Aether expressing Fibonacci and Lucas Sequences joined at the hip: a thought experiment

For "r" = 0

n	F(n)	F(n) * (-1) * F(-n)	F(n-1) * F(-(n+1))	F(n-2) * (-1) * F(-(n+2))	F(n-3) * F(-(n+3))	F(n-4) * (-1) * F(-(n+4))	F(n-5) * F(-(n+5))
-3	5	0	17	-3	26	-195	145
-2	-2	0	17	-3	26	-195	145
-1	3	0	17	-3	26	-195	145
0	1	0	17	-3	26	-195	145
1	4	0	17	-3	26	-195	145
2	5	0	17	-3	26	-195	145
3	9	0	17	-3	26	-195	145

For "r" = 1

n	F(n)	F(n) * (-1) * F(-n)	F(n-1) * F(-(n+1))	F(n-2) * (-1) * F(-(n+2))	F(n-3) * F(-(n+3))	F(n-4) * (-1) * F(-(n+4))	F(n-5) * F(-(n+5))
-3	5	-17	0	-20	9	-212	128
-2	-2	-17	0	-20	9	-212	128
-1	3	-17	0	-20	9	-212	128
0	1	-17	0	-20	9	-212	128
1	4	-17	0	-20	9	-212	128
2	5	-17	0	-20	9	-212	128
3	9	-17	0	-20	9	-212	128

For "r" = 2

n	F(n)	F(n) * (-1) * F(-n)	F(n-1) * F(-(n+1))	F(n-2) * (-1) * F(-(n+2))	F(n-3) * F(-(n+3))	F(n-4) * (-1) * F(-(n+4))	F(n-5) * F(-(n+5))
-3	5	3	20	0	29	-192	148
-2	-2	3	20	0	29	-192	148
-1	3	3	20	0	29	-192	148
0	1	3	20	0	29	-192	148
1	4	3	20	0	29	-192	148
2	5	3	20	0	29	-192	148
3	9	3	20	0	29	-192	148

An Aether expressing Fibonacci and Lucas Sequences joined at the hip: a thought experiment

For "r" = 3

n	F(n)	F(n) * (-1) * F(-n)	F(n-1) * F(-(n+1))	F(n-2) * (-1) * F(-(n+2))	F(n-3) * F(-(n+3))	F(n-4) * (-1) * F(-(n+4))	F(n-5) * F(-(n+5))
-3	5	-26	-9	-29	0	-221	119
-2	-2	-26	-9	-29	0	-221	119
-1	3	-26	-9	-29	0	-221	119
0	1	-26	-9	-29	0	-221	119
1	4	-26	-9	-29	0	-221	119
2	5	-26	-9	-29	0	-221	119
3	9	-26	-9	-29	0	-221	119

For "r" = 4

n	F(n)	F(n) * (-1) * F(-n)	F(n-1) * F(-(n+1))	F(n-2) * (-1) * F(-(n+2))	F(n-3) * F(-(n+3))	F(n-4) * (-1) * F(-(n+4))	F(n-5) * F(-(n+5))
-3	5	195	212	192	221	0	340
-2	-2	195	212	192	221	0	340
-1	3	195	212	192	221	0	340
0	1	195	212	192	221	0	340
1	4	195	212	192	221	0	340
2	5	195	212	192	221	0	340
3	9	195	212	192	221	0	340

For "r" = 5

n	F(n)	F(n) * (-1) * F(-n)	F(n-1) * F(-(n+1))	F(n-2) * (-1) * F(-(n+2))	F(n-3) * F(-(n+3))	F(n-4) * (-1) * F(-(n+4))	F(n-5) * F(-(n+5))
-3	5	-145	-128	-148	-119	-340	0
-2	-2	-145	-128	-148	-119	-340	0
-1	3	-145	-128	-148	-119	-340	0
0	1	-145	-128	-148	-119	-340	0
1	4	-145	-128	-148	-119	-340	0
2	5	-145	-128	-148	-119	-340	0
3	9	-145	-128	-148	-119	-340	0

An Aether expressing Fibonacci and Lucas Sequences joined at the hip: a thought experiment

Tables of harmonic Fibonacci and Lucas sequences

F(n) = F(n+1) - F(n-1)	L(n) = F(n+1) + F(n-1)	U(n-) = F(n+1) - F(n-2)	U(n+) = F(n+1) + F(n-2)
34	-76	68	68
-21	47	-42	-42
13	-29	26	26
-8	18	-16	-16
5	-11	10	10
-3	7	-6	-6
2	-4	4	4
-1	3	-2	-2
1	-1	2	2
0	2	0	0
1	1	2	2
1	3	2	2
2	4	4	4
3	7	6	6
5	11	10	10
8	18	16	16
13	29	26	26
21	47	42	42
34	76	68	68

Cassini Idenities	1		-5	Cassini Idenities	4	4
Average of Cassini Values	3	L(2) or F(4)		Average of Cassini Values	4	L(3)
Difference in Cassini Values		4		Difference in Cassini Values		0

An Aether expressing Fibonacci and Lucas Sequences joined at the hip: a thought experiment

U(n-) = F(n+1) - F(n-3)	U(n+) = F(n+1) + F(n-3)	U(n-) = F(n+1) - F(n-4)	U(n+) = F(n+1) + F(n-4)
-76	102	157	81
47	-63	-97	-50
-29	39	60	31
18	-24	-37	-19
-11	15	23	12
7	-9	-14	-7
-4	6	9	5
3	-3	-5	-2
-1	3	4	3
2	0	-1	1
1	3	3	4
3	3	2	5
4	6	5	9
7	9	7	14
11	15	12	23
18	24	19	37
29	39	31	60
47	63	50	97
76	102	81	157

Cassini Idenities	-5		9	Cassini Idenities	11		11
Average of Cassini Values	7	L(4)		Average of Cassini Values	11	L(5)	
Difference in Cassini Values		4		Difference in Cassini Values		0	

237

An Aether expressing Fibonacci and Lucas Sequences joined at the hip: a thought experiment

U(n-) = F(n+1) - F(n-5)	U(n+) = F(n+1) + F(n-5)		U(n-) = F(n+1) - F(n-6)	U(n+) = F(n+1) + F(n-6)	
136	-152		149	225	
-84	94		-92	-139	
52	-58		57	86	
-32	36		-35	-53	
20	-22		22	33	
-12	14		-13	-20	
8	-8		9	13	
-4	6		-4	-7	
4	-2		5	6	
0	4		1	-1	
4	2		6	5	
4	6		7	4	
8	8		13	9	
12	14		20	13	
20	22		33	22	
32	36		53	35	
52	58		86	57	
84	94		139	92	
136	152		225	149	
Cassini Idenities	16	-20	Cassini Idenities	29	29
Average of Cassini Values	18	L(6)	Average of Cassini Values	29	L(7)
Difference in Cassini Values	4		Difference in Cassini Values	0	

238

An Aether expressing Fibonacci and Lucas Sequences joined at the hip: a thought experiment

U(n-) = F(n+1) - F(n-7)	U(n+) = F(n+1) + F(n-7)	U(n-) = F(n+1) - F(n-8)	U(n+) = F(n+1) + F(n-8)
-228	238	382	230
141	-147	-236	-142
-87	91	146	88
54	-56	-90	-54
-33	35	56	34
21	-21	-34	-20
-12	14	22	14
9	-7	-12	-6
-3	7	10	8
6	0	-2	2
3	7	8	10
9	7	6	12
12	14	14	22
21	21	20	34
33	35	34	56
54	56	54	90
87	91	88	146
141	147	142	236
228	238	230	382
Cassini Idenities -45	49	Cassini Idenities 76	76
Average of Cassini Values 47	L(8)	Average of Cassini Values 76	L(9)
Difference in Cassini Values	4	Difference in Cassini Values	0

An Aether expressing Fibonacci and Lucas Sequences joined at the hip: a thought experiment

U(n-) = F(n+1) - F(n-9)	U(n+) = F(n+1) + F(n-9)		U(n-) = F(n+1) - F(n-10)	U(n+) = F(n+1) + F(n-10)	
374	-380		379	607	
-231	235		-234	-375	
143	-145		145	232	
-88	90		-89	-143	
55	-55		56	89	
-33	35		-33	-54	
22	-20		23	35	
-11	15		-10	-19	
11	-5		13	16	
0	10		3	-3	
11	5		16	13	
11	15		19	10	
22	20		35	23	
33	35		54	33	
55	55		89	56	
88	90		143	89	
143	145		232	145	
231	235		375	234	
374	380		607	379	
Cassini Idenities	121	-125	Cassini Idenities	199	199
Average of Cassini Values	123	L(10)	Average of Cassini Values	199	L(11)
Difference in Cassini Values	4		Difference in Cassini Values	0	

An Aether expressing Fibonacci and Lucas Sequences joined at the hip: a thought experiment

U(n-) = F(n+1) - F(n-11)	U(n+) = F(n+1) + F(n-11)		U(n-) = F(n+1) - F(n-12)	U(n+) = F(n+1) + F(n-12)
-608	612		989	609
376	-378		-611	-376
-232	234		378	233
144	-144		-233	-143
-88	90		145	90
56	-54		-88	-53
-32	36		57	37
24	-18		-31	-16
-8	18		26	21
16	0		-5	5
8	18		21	26
24	18		16	31
32	36		37	57
56	54		53	88
88	90		90	145
144	144		143	233
232	234		233	378
376	378		376	611
608	612		609	989

ssini Idenities	-320		324	Cassini Idenities	521		521
Average of Cassini Values	322	L(12)		Average of Cassini Values	521	L(13)	
Difference in Cassini Values		4		Difference in Cassini Values		0	

241

An Aether expressing Fibonacci and Lucas Sequences joined at the hip: a thought experiment

U(n-) = F(n+1) - F(n-13)		U(n+) = F(n+1) + F(n-13)		U(n-) = F(n+1) - F(n-14)		U(n+) = F(n+1) + F(n-14)	
986		-988		988		1596	
-609		611		-610		-986	
377		-377		378		610	
-232		234		-232		-376	
145		-143		146		234	
-87		91		-86		-142	
58		-52		60		92	
-29		39		-26		-50	
29		-13		34		42	
0		26		8		-8	
29		13		42		34	
29		39		50		26	
58		52		92		60	
87		91		142		86	
145		143		234		146	
232		234		376		232	
377		377		610		378	
609		611		986		610	
986		988		1596		988	
Cassini Idenities	841		-845	Cassini Idenities	1364		1364
Average of Cassini Values	843	L(14)		Average of Cassini Values	1364	L(15)	
Difference in Cassini Values		4		Difference in Cassini Values		0	

An Aether expressing Fibonacci and Lucas Sequences joined at the hip: a thought experiment

U(n-) = F(n+1) - F(n-15)	U(n+) = F(n+1) + F(n-15)		U(n-) = F(n+1) - F(n-16)	U(n+) = F(n+1) + F(n-16)
-1596	1598		2585	1597
987	-987		-1597	-986
-609	611		988	611
378	-376		-609	-375
-231	235		379	236
147	-141		-230	-139
-84	94		149	97
63	-47		-81	-42
-21	47		68	55
42	0		-13	13
21	47		55	68
63	47		42	81
84	94		97	149
147	141		139	230
231	235		236	379
378	376		375	609
609	611		611	988
987	987		986	1597
1596	1598		1597	2585

es -2205		2209	Cassini Idenities 3571		3571
Average of Cassini Values 2207	L(16)		Average of Cassini Values 3571	L(17)
Difference in Cassini Values	4		Difference in Cassini Values	0

An Aether expressing Fibonacci and Lucas Sequences joined at the hip: a thought experiment

U(n-) = F(n+1) - F(n-17)	U(n+) = F(n+1) + F(n-17)		U(n-) = F(n+1) - F(n-18)	U(n+) = F(n+1) + F(n-18)	
2584	-2584		2585	4181	
-1596	1598		-1596	-2583	
988	-986		989	1598	
-608	612		-607	-985	
380	-374		382	613	
-228	238		-225	-372	
152	-136		157	241	
-76	102		-68	-131	
76	-34		89	110	
0	68		21	-21	
76	34		110	89	
76	102		131	68	
152	136		241	157	
228	238		372	225	
380	374		613	382	
608	612		985	607	
988	986		1598	989	
1596	1598		2583	1596	
2584	2584		4181	2585	
Cassini Idenities 5776		-5780	Cassini Idenities 9349		9349
Average of Cassini Values 5778	L(18)		Average of Cassini Values 9349	L(19)	
Difference in Cassini Values	4		Difference in Cassini Values	0	

An Aether expressing Fibonacci and Lucas Sequences joined at the hip: a thought experiment

U(n-) = F(n+1) - F(n-19)	U(n+) = F(n+1) + F(n-19)	U(n-) = F(n+1) - F(n-20)	U(n+) = F(n+1) + F(n-20)
-4180	4182	6766	4182
2585	-2583	-4180	-2582
-1595	1599	2586	1600
990	-984	-1594	-982
-605	615	992	618
385	-369	-602	-364
-220	246	390	254
165	-123	-212	-110
-55	123	178	144
110	0	-34	34
55	123	144	178
165	123	110	212
220	246	254	390
385	369	364	602
605	615	618	992
990	984	982	1594
1595	1599	1600	2586
2585	2583	2582	4180
4180	4182	4182	6766

Cassini Idenities	-15125		15129	Cassini Idenities	24476		24476
Average of Cassini Values	15127	L(20)		Average of Cassini Values	24476	L(21)	
Difference in Cassini Values		4		Difference in Cassini Values		0	

An Aether expressing Fibonacci and Lucas Sequences joined at the hip: a thought experiment

Spin correction tables for Fibonacci Supra-MNS states

Tables for MNS-F(4)^2

Standard Supra-MNS states				Word Index n	Spin Corrected		
Words	Sum (x+)		Sum (x-)		Sum (x+)		Sum (x-)
		0				0	
Fr -Sum 1	4		2	1	4		2
Fr -Sum 2	8		1	2	-1		1
Fr -Sum 3	7		5	3	-2		-4
Fr -Sum 4	5		7	4	-4		-2
Fr -Sum 5	1		8	5	1		-1
Fr -Sum 6	2		4	6	2		4

An Aether expressing Fibonacci and Lucas Sequences joined at the hip: a thought experiment

Tables for MNS-F(5)^2

Standard Supra-MNS states			Word Index n	Spin Corrected		
Words	Sum (x+)	Sum (x-)		Sum (x+)		Sum (x-)
		0			0	
Fr -Sum 1	7	24	1	7		-1
Fr -Sum 2	1	18	2	1		-7
Fr -Sum 3	18	1	3	-7		1
Fr -Sum 4	24	7	4	-1		7
Fr -Sum 5	7	24	5	7		-1
Fr -Sum 6	1	18	6	1		-7
Fr -Sum 7	18	1	7	-7		1
Fr -Sum 8	24	7	8	-1		7
Fr -Sum 9	7	24	9	7		-1
Fr -Sum 10	1	18	10	1		-7
Fr -Sum 11	18	1	11	-7		1
Fr -Sum 12	24	7	12	-1		7
Fr -Sum 13	7	24	13	7		-1
Fr -Sum 14	1	18	14	1		-7
Fr -Sum 15	18	1	15	-7		1
Fr -Sum 16	24	7	16	-1		7
Fr -Sum 17	7	24	17	7		-1
Fr -Sum 18	1	18	18	1		-7
Fr -Sum 19	18	1	19	-7		1
Fr -Sum 20	24	7	20	-1		7

An Aether expressing Fibonacci and Lucas Sequences joined at the hip: a thought experiment

Tables for MNS-F(6)^2

Standard Supra-MNS states				Word Index n	Spin Corrected		
Words	Sum (x+)		Sum (x-)		Sum (x+)		Sum (x-)
		0				0	
Fr -Sum 1	12		4	1	12		4
Fr -Sum 2	20		60	2	20		-4
Fr -Sum 3	28		52	3	28		-12
Fr -Sum 4	36		44	4	-28		-20
Fr -Sum 5	44		36	5	-20		-28
Fr -Sum 6	52		28	6	-12		28
Fr -Sum 7	60		20	7	-4		20
Fr -Sum 8	4		12	8	4		12
Fr -Sum 9	12		4	9	12		4
Fr -Sum 10	20		60	10	20		-4
Fr -Sum 11	28		52	11	28		-12
Fr -Sum 12	36		44	12	-28		-20
Fr -Sum 13	44		36	13	-20		-28
Fr -Sum 14	52		28	14	-12		28
Fr -Sum 15	60		20	15	-4		20
Fr -Sum 16	4		12	16	4		12

An Aether expressing Fibonacci and Lucas Sequences joined at the hip: a thought experiment

Tables for MNS-F(7)^2

Standard Supra-MNS states			Word Index n	Spin Corrected		
Words	Sum (x+)		Sum (x-)		Sum (x+)	Sum (x-)
		0				0
Fr -Sum 1	20		165	1	20	-4
Fr -Sum 2	69		136	2	69	-33
Fr -Sum 3	162		108	3	-7	-61
Fr -Sum 4	35		46	4	35	46
Fr -Sum 5	163		126	5	-6	-43
Fr -Sum 6	30		110	6	30	-59
Fr -Sum 7	19		147	7	19	-22
Fr -Sum 8	74		72	8	74	72
Fr -Sum 9	137		87	9	-32	-82
Fr -Sum 10	160		84	10	-9	84
Fr -Sum 11	45		17	11	45	17
Fr -Sum 12	113		98	12	-56	-71
Fr -Sum 13	111		48	13	-58	48
Fr -Sum 14	121		58	14	-48	58
Fr -Sum 15	71		56	15	71	56
Fr -Sum 16	152		124	16	-17	-45
Fr -Sum 17	85		9	17	-84	9
Fr -Sum 18	82		32	18	82	32
Fr -Sum 19	97		95	19	-72	-74
Fr -Sum 20	22		150	20	22	-19
Fr -Sum 21	59		139	21	59	-30
Fr -Sum 22	43		6	22	43	6
Fr -Sum 23	123		134	23	-46	-35
Fr -Sum 24	61		7	24	61	7
Fr -Sum 25	33		100	25	33	-69
Fr -Sum 26	4		149	26	4	-20

An Aether expressing Fibonacci and Lucas Sequences joined at the hip: a thought experiment

Standard Supra-MNS states				Word Index n	Spin Corrected		
Words	Sum (x+)		Sum (x-)		Sum (x+)		Sum (x-)
		0				0	
Fr -Sum 27	149		4	27	-20		4
Fr -Sum 28	100		33	28	-69		33
Fr -Sum 29	7		61	29	7		61
Fr -Sum 30	134		123	30	-35		-46
Fr -Sum 31	6		43	31	6		43
Fr -Sum 32	139		59	32	-30		59
Fr -Sum 33	150		22	33	-19		22
Fr -Sum 34	95		97	34	-74		-72
Fr -Sum 35	32		82	35	32		82
Fr -Sum 36	9		85	36	9		-84
Fr -Sum 37	124		152	37	-45		-17
Fr -Sum 38	56		71	38	56		71
Fr -Sum 39	58		121	39	58		-48
Fr -Sum 40	48		111	40	48		-58
Fr -Sum 41	98		113	41	-71		-56
Fr -Sum 42	17		45	42	17		45
Fr -Sum 43	84		160	43	84		-9
Fr -Sum 44	87		137	44	-82		-32
Fr -Sum 45	72		74	45	72		74
Fr -Sum 46	147		19	46	-22		19
Fr -Sum 47	110		30	47	-59		30
Fr -Sum 48	126		163	48	-43		-6
Fr -Sum 49	46		35	49	46		35
Fr -Sum 50	108		162	50	-61		-7
Fr -Sum 51	136		69	51	-33		69
Fr -Sum 52	165		20	52	-4		20

Epilogue 10:

Lucas values and tables

Tables for MNS-F(4)^2

Standard				n	Spin Corrected			
F(r)		F(-r)	F(r) * F(-r)		F(r)		F(-r)	F(r) * F(-r)
2			4		2			4	
1		-1	-1	1	1		-1	-1	
3		3	0	2	3		3	0	I
4		-4	2	3	4		-4	2	*
7		7	4	4	-2		-2	4	*
2		-2	-4	5	2		-2	-4	
0		0	0	6	0		0	0	I I
2		-2	-4	7	2		-2	-4	
2		2	4	8	2		2	4	*
4		-4	2	9	4		-4	2	*
6		6	0	10	-3		-3	0	I
1		-1	-1	11	1		-1	-1	
7		7	4	12	-2		-2	4	*
8		-8	8	13	-1		1	-1	
6		6	0	14	-3		-3	0	I
5		-5	2	15	-4		4	2	
2		2	4	16	2		2	4	*
7		-7	5	17	-2		2	-4	
0		0	0	18	0		0	0	I I
7		-7	5	19	-2		2	-4	
7		7	4	20	-2		-2	4	*
5		-5	2	21	-4		4	2	
3		3	0	22	3		3	0	I
8		-8	8	23	-1		1	-1	
2		2	4	24	2		2	4	*

An Aether expressing Fibonacci and Lucas Sequences joined at the hip: a thought experiment

Tables of words using "4" as punctuation (word = 3 bits length).

Carrier states				Supra-MNS states		
Original Sum		Corrected Sum	Word Index n	Sum (r+)		Sum (r-)
	4				4	
1		1	1	8		7
-8		1	2	4		5
1		1	3	2		1
1		1	4	1		2
-8		1	5	5		4
1		1	6	7		8

Standard Supra-MNS states					Spin Corrected		
Words	Sum (r+)		Sum (r-)	Word Index n	Sum (r+)		Sum (r-)
		4				4	
Fr -Sum 1	8		7	1	-1		-2
Fr -Sum 2	4		5	2	4		-4
Fr -Sum 3	2		1	3	2		1
Fr -Sum 4	1		2	4	1		2
Fr -Sum 5	5		4	5	-4		4
Fr -Sum 6	7		8	6	-2		-1

An Aether expressing Fibonacci and Lucas Sequences joined at the hip: a thought experiment

Tables of words using "0" as punctuation (word = 3 bits length).

Carrier states			Word Index n	Supra-MNS states		
Original Sum		Corrected Sum		Sum (r+)		Sum (r-)
	4				4	
2		2	1	2		2
2		2	2	4		1
2		2	3	8		5
2		2	4	7		7
2		2	5	5		8
2		2	6	1		4
2		2	7	2		2

Standard Supra-MNS states				Word Index n	Spin Corrected		
Words	Sum (r+)		Sum (r-)		Sum (r+)		Sum (r-)
		4				4	
Fr -Sum 1	2		2	1	2		2
Fr -Sum 2	4		1	2	4		1
Fr -Sum 3	8		5	3	-1		-4
Fr -Sum 4	7		7	4	-2		-2
Fr -Sum 5	5		8	5	-4		-1
Fr -Sum 6	1		4	6	1		4
Fr -Sum 7	2		2	7	2		2

An Aether expressing Fibonacci and Lucas Sequences joined at the hip: a thought experiment

Notes on the Carrier and Supra-MNS states involving the Lucas sequence.

To noted here is that in using "0" as a punctuation marker in the Lucas sequence, there arises a situation at:

- an inflection point at "Fr-sum 1"; and at
- a point in each time-strand at "Fr-sum 7" [and at every further 6 time points in the forward and revers time strands]

that continuity of the 1_2_4_8_7_5 Aether-state cycle is maintained between the two strands only if it is assumed that the "2" state value is somehow shared between the two "time-strands" at those points. I don´t know if this provides any insight to an enabling mechanism for the phenomena of "deja-vu", but I leave this possibility to the imagination of any budding science fiction writer reading these words. The plotline of "Tenet" by Christopher Nolan is a good attempt though.

Another item to be noted from the various Supra-MNS states is that, if one multiplies corresponding Supra-MNS states of the different time strands with one another for different punctuation paradigms, then:

- with the Fibonacci sequence using "0" as punctuation, a product having an Aether value of "4";

An Aether expressing Fibonacci and Lucas Sequences joined at the hip: a thought experiment

- with the Lucas sequence using "4" as punctuation, a product having an Aether value of "1"; and
- with the Lucas sequence using "0" as punctuation, a product having an Aether value of "2".

An Aether expressing Fibonacci and Lucas Sequences joined at the hip: a thought experiment

Tables for MNS-F(5)^2

Standard				n	Spin Corrected				
F(r)		F(-r)	F(r) * F(-r)	n	F(r)		F(-r)	F(r) * F(-r)	
2			4			2		4	
1		-1	-1	1	1		-1	-1	
3		3	9	2	3		3	9	
4		-4	-16	3	4		-4	9	*
7		7	24	4	7		7	-1	
11		-11	4	5	11		-11	4	*
18		18	24	6	-7		-7	-1	
4		-4	-16	7	4		-4	9	*
22		22	9	8	-3		-3	9	
1		-1	-1	9	1		-1	-1	
23		23	4	10	-2		-2	4	*
24		-24	24	11	-1		1	-1	
22		22	9	12	-3		-3	9	
21		-21	9	13	-4		4	9	
18		18	24	14	-7		-7	-1	
14		-14	4	15	-11		11	4	*
7		7	24	16	7		7	-1	
21		-21	9	17	-4		4	9	
3		3	9	18	3		3	9	
24		-24	24	19	-1		1	-1	
2		2	4	20	2		2	4	*
1		-1	-1	21	1		-1	-1	
3		3	9	22	3		3	9	
4		-4	-16	23	4		-4	9	*
7		7	24	24	7		7	-1	
11		-11	4	25	11		-11	4	*

The Lucas sequence with the MNS-25 circuit requires 100 iterations before the cycle begins again. For efficiency, only the first 25 are shown here.

An Aether expressing Fibonacci and Lucas Sequences joined at the hip: a thought experiment

Tables of words using "4" as punctuation (word = 4 bits length).

Carrier states				Supra-MNS states		
Original Sum		Corrected Sum	Word Index n	Sum (r+)		Sum (r-)
	4				4	
16		-9	1	15		5
16		-9	2	20		10
16		-9	3	10		20
16		-9	4	5		15
16		-9	5	15		5
16		-9	6	20		10
16		-9	7	10		20
16		-9	8	5		15
16		-9	9	15		5
16		-9	10	20		10
16		-9	11	10		20
16		-9	12	5		15
16		-9	13	15		5
16		-9	14	20		10
16		-9	15	10		20
16		-9	16	5		15
16		-9	17	15		5
16		-9	18	20		10
16		-9	19	10		20
16		-9	20	5		15

An Aether expressing Fibonacci and Lucas Sequences joined at the hip: a thought experiment

Standard Supra-MNS states				Word Index n	Spin Corrected		
Words	Sum (r+)		Sum (r-)		Sum (r+)		Sum (r-)
		4				4	
Fr -Sum 1	15		5	1	-10		5
Fr -Sum 2	20		10	2	-5		10
Fr -Sum 3	10		20	3	10		-5
Fr -Sum 4	5		15	4	5		-10
Fr -Sum 5	15		5	5	-10		5
Fr -Sum 6	20		10	6	-5		10
Fr -Sum 7	10		20	7	10		-5
Fr -Sum 8	5		15	8	5		-10
Fr -Sum 9	15		5	9	-10		5
Fr -Sum 10	20		10	10	-5		10
Fr -Sum 11	10		20	11	10		-5
Fr -Sum 12	5		15	12	5		-10
Fr -Sum 13	15		5	13	-10		5
Fr -Sum 14	20		10	14	-5		10
Fr -Sum 15	10		20	15	10		-5
Fr -Sum 16	5		15	16	5		-10
Fr -Sum 17	15		5	17	-10		5
Fr -Sum 18	20		10	18	-5		10
Fr -Sum 19	10		20	19	10		-5
Fr -Sum 20	5		15	20	5		-10

An Aether expressing Fibonacci and Lucas Sequences joined at the hip: a thought experiment

Tables of words using "0" as punctuation (word = 4 bits length).

There are no words using "0" as punctuation. In MNS circuits using the Lucas sequence, only MNS-L(n)^2 circuits will give rise to "words" using "0" as punctuation, i.e., as well as using "4" as punctuation.

An Aether expressing Fibonacci and Lucas Sequences joined at the hip: a thought experiment

Notes on the Carrier and Supra-MNS states involving the Lucas sequence for MNS-F(5)^2 circuits.

Here the carrier states may be considered to have a value of either L(3)^2 [=16] or − F(4)^2 [=-9]. It is noted that if one adds the value of the carrier state for the Fibonacci sequences [= -5 or 20] to one or the other of the carrier states for the Lucas sequence, one gets:

- 11 [= L(5)] and
- -4 [= L(-3)]; or
- 11 [= 36 mod 25] [= L(5)]; and
- 11 [= L(5)].

The Supra-MNS states involves a recurring sequence of:

- 5 _ 15 _ 20 _10

which can be broken down into F(5) [= 5] times the recurring sequence

- 1 _ 3 _ 4 _ 2 .

This recurring sequence *per se* gives arises from a Rodin type force graph involving Modulo-5 as an attractor with mutually complementary multipliers of:

- 2 [= L(0) or F(3)] and 3 [= L(2) or F(4)].

An Aether expressing Fibonacci and Lucas Sequences joined at the hip: a thought experiment

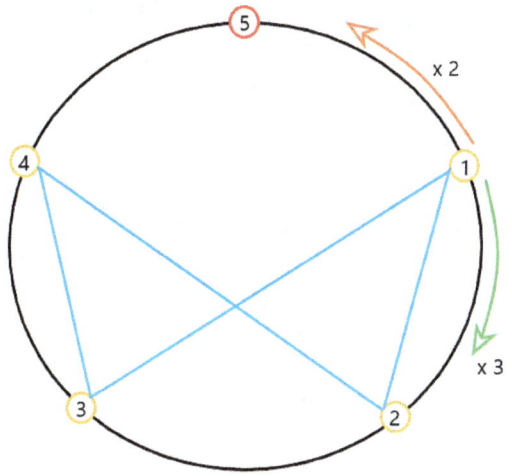

Rodin-type force graph for Modulo-5 with multipliers "2" and "3"

An Aether expressing Fibonacci and Lucas Sequences joined at the hip: a thought experiment

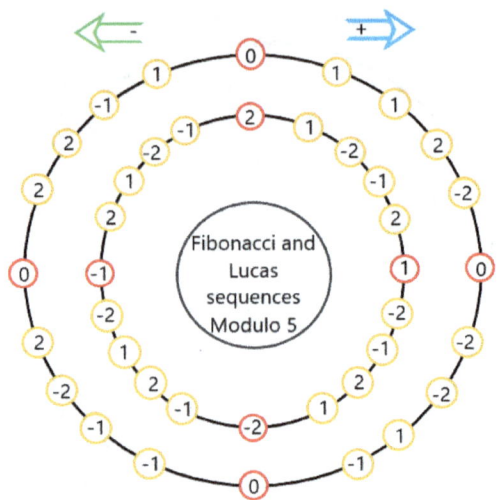

20 Aether-elements with modulo-5 of each of the Fibonacci sequence and the Lucas sequence

An Aether expressing Fibonacci and Lucas Sequences joined at the hip: a thought experiment

What we see here, with the Lucas sequence in modulo 25, is that there is a correlation between the words of the Supra-MNS values of the Lucas sequence and the cycle shown by multipliers of Catalan Identities of "13" and "2".

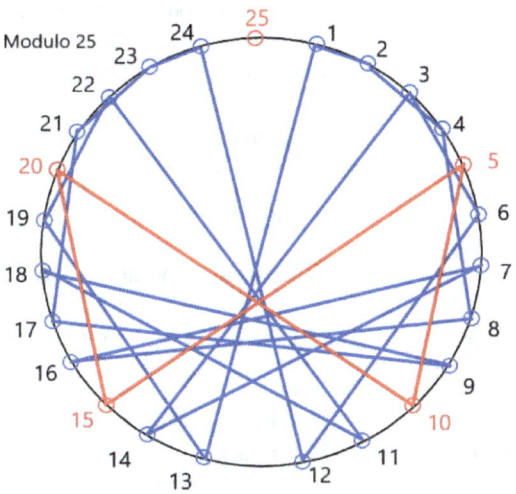

As an observation, I submit we see here a more fundamental internal dynamism in the driving of the MNS-$F(5)^2$ structure when employing the Cassini Identity of "5" for the Lucas sequence with the Aether elements of the Rodin force graph of the MNS-5 sequence. if we were trying to find an atomic structure involving 13 x 2 electrons in its outer shells and conversely 26 protons in its nucleus, we would look to Iron. Is it a coincidence that iron is associated with the force of magnetism?

263

An Aether expressing Fibonacci and Lucas Sequences joined at the hip: a thought experiment

Another observation is that the Cassini identity of "5" of the Lucas sequence per se is itself appears to be used commonly as a multiplier in both:

- The Rodin force graph developed from the Supra-MNS words of the Lucas sequence in the MNS-$F(5)^2$ circuit; and
- The Cassini multiplier opposite that of "2" of the Rodin force graph developed from the Supra-MNS words of both the Fibonacci sequence and the Lucas sequence in the MNS-$F(4)^2$ circuit.

To put this out there, if we treat:

- the MNS-$F(4)^2$ circuit as being directed to electrical current, and
- the MNS-$F(5)^2$ circuit as being directed to magnetism,

then can we treat the Cassini Identity of "5" for the Lucas sequence as being the connection between the MNS circuits which enables electro-magnetism in the physical world?

An Aether expressing Fibonacci and Lucas Sequences joined at the hip: a thought experiment

Tables for MNS-F$(7)^2$

Standard				n	Spin Corrected			
F(r)		F(-r)	F(r) * F(-r)		F(r)		F(-r)	F(r) * F(-r)
	2		4			2		4
1		-1	-1	1	1		-1	-1
3		3	9	2	3		3	9
4		-4	-16	3	4		-4	-16
7		7	49	4	7		7	49
11		-11	-121	5	11		-11	48
18		18	155	6	18		18	-14
29		-29	4	7	29		-29	4
47		47	12	8	47		47	12
76		-76	139	9	76		-76	-30
123		123	88	10	-46		-46	-81
30		-30	114	11	30		-30	-55
153		153	87	12	-16		-16	-82
14		-14	142	13	14		-14	-27
167		167	4	14	-2		-2	4
12		-12	-144	15	12		-12	25
10		10	100	16	10		10	-69
22		-22	23	17	22		-22	23
32		32	10	18	32		32	10
54		-54	126	19	54		-54	-43
86		86	129	20	-83		-83	-40
140		-140	4	21	-29		29	4
57		57	38	22	57		57	38
28		-28	61	23	28		-28	61
85		85	127	24	-84		-84	-42
113		-113	75	25	-56		56	75
29		29	165	26	29		29	-4
142		-142	116	27	-27		27	-53
2		2	4	28	2		2	4

The Lucas sequence with the MNS-169 circuit requires 364 iterations before the cycle begins again. For efficiency, only the first 28 are shown here.

An Aether expressing Fibonacci and Lucas Sequences joined at the hip: a thought experiment

Carrier states				Supra-MNS states		
Original Sum		Corrected Sum	Word Index n	Sum (r+)		Sum (r-)
	4				4	
75		75	1	44		12
-263		75	2	105		34
-94		75	3	47		40
75		75	4	116		57
75		75	5	31		77
75		75	6	1		21
-94		75	7	60		144
75		75	8	51		70
75		75	9	18		142
75		75	10	66		8
-94		75	11	73		79
-263		75	12	155		83
75		75	13	5		38
75		75	14	131		164
-263		75	15	86		14
-94		75	16	90		96
75		75	17	161		103
75		75	18	27		151
75		75	19	99		118
-94		75	20	25		109
75		75	21	148		168
75		75	22	92		138
75		75	23	112		53
-94		75	24	129		122
-263		75	25	135		64
75		75	26	157		125

An Aether expressing Fibonacci and Lucas Sequences joined at the hip: a thought experiment

Carrier states				Supra-MNS states		
Original Sum		Corrected Sum	Word Index n	Sum (r+)		Sum (r-)
	4				4	
75		75	27	125		157
-263		75	28	64		135
-94		75	29	122		129
75		75	30	53		112
75		75	31	138		92
75		75	32	168		148
-94		75	33	109		25
75		75	34	118		99
75		75	35	151		27
75		75	36	103		161
-94		75	37	96		90
-263		75	38	14		86
75		75	39	164		131
75		75	40	38		5
-263		75	41	83		155
-94		75	42	79		73
75		75	43	8		66
75		75	44	142		18
75		75	45	70		51
-94		75	46	144		60
75		75	47	21		1
75		75	48	77		31
75		75	49	57		116
-94		75	50	40		47
-263		75	51	34		105
75		75	52	12		44

An Aether expressing Fibonacci and Lucas Sequences joined at the hip: a thought experiment

Standard Supra-MNS states			Word Index n	Spin Corrected		
Words	Sum (r+)	Sum (r-)		Sum (r+)		Sum (r-)
		4			4	
Fr -Sum 1	44	12	1	44		12
Fr -Sum 2	105	34	2	-64		34
Fr -Sum 3	47	40	3	47		40
Fr -Sum 4	116	57	4	-53		57
Fr -Sum 5	31	77	5	31		77
Fr -Sum 6	1	21	6	1		21
Fr -Sum 7	60	144	7	60		-25
Fr -Sum 8	51	70	8	51		70
Fr -Sum 9	18	142	9	18		-27
Fr -Sum 10	66	8	10	66		8
Fr -Sum 11	73	79	11	73		79
Fr -Sum 12	155	83	12	-14		83
Fr -Sum 13	5	38	13	5		38
Fr -Sum 14	131	164	14	-38		-5
Fr -Sum 15	86	14	15	-83		14
Fr -Sum 16	90	96	16	-79		-73
Fr -Sum 17	161	103	17	-8		-66
Fr -Sum 18	27	151	18	27		-18
Fr -Sum 19	99	118	19	-70		-51
Fr -Sum 20	25	109	20	25		-60
Fr -Sum 21	148	168	21	-21		-1
Fr -Sum 22	92	138	22	-77		-31
Fr -Sum 23	112	53	23	-57		53
Fr -Sum 24	129	122	24	-40		-47
Fr -Sum 25	135	64	25	-34		64
Fr -Sum 26	157	125	26	-12		-44

An Aether expressing Fibonacci and Lucas Sequences joined at the hip: a thought experiment

Standard Supra-MNS states				Word Index n	Spin Corrected		
Words	Sum (r+)		Sum (r-)		Sum (r+)		Sum (r-)
		4				4	
Fr -Sum 27	125		157	27	-44		-12
Fr -Sum 28	64		135	28	64		-34
Fr -Sum 29	122		129	29	-47		-40
Fr -Sum 30	53		112	30	53		-57
Fr -Sum 31	138		92	31	-31		-77
Fr -Sum 32	168		148	32	-1		-21
Fr -Sum 33	109		25	33	-60		25
Fr -Sum 34	118		99	34	-51		-70
Fr -Sum 35	151		27	35	-18		27
Fr -Sum 36	103		161	36	-66		-8
Fr -Sum 37	96		90	37	-73		-79
Fr -Sum 38	14		86	38	14		-83
Fr -Sum 39	164		131	39	-5		-38
Fr -Sum 40	38		5	40	38		5
Fr -Sum 41	83		155	41	83		-14
Fr -Sum 42	79		73	42	79		73
Fr -Sum 43	8		66	43	8		66
Fr -Sum 44	142		18	44	-27		18
Fr -Sum 45	70		51	45	70		51
Fr -Sum 46	144		60	46	-25		60
Fr -Sum 47	21		1	47	21		1
Fr -Sum 48	77		31	48	77		31
Fr -Sum 49	57		116	49	57		-53
Fr -Sum 50	40		47	50	40		47
Fr -Sum 51	34		105	51	34		-64
Fr -Sum 52	12		44	52	12		44

An Aether expressing Fibonacci and Lucas Sequences joined at the hip: a thought experiment

Summary of Lucas Carrier and supra-MNS states for the MNS-F(7)^2 circuit

Carrier states and Supra-MNS states for each of the 52 words from the 364 iterations of a full cycle.

- Non-spin corrected carrier states oscillate between "75" and "-94"; and
- Supra-MNS states are generated by multipliers "60" and "31".

As an observation regarding the value "31":

- "31" is the sum of L(7) [= 29] and L(0) [= 2]
- "60" –"31" = L(7) [= 29]
- If we compare carrier states of the Fibonacci sequence (125 and -44) to those of the Lucas sequence (75 and -94) for MNS-F(7)^2 circuits, we see that they are different from one another:
    - in one aspect by a value of "50"
        - (125 -75 = 50 and -44 – (-94) = 50),

    but

    - in another aspect by a value of "31"
        - (125 + (-94) = 31 and 75 + (-44) = 31)
- The value "31" is itself a harmonic of the Cassini Identity derived from the unbalanced sequence:
    - -34 _ 23 _ -11 _ 12 _ 1 _ 13 _ 14

An Aether expressing Fibonacci and Lucas Sequences joined at the hip: a thought experiment

That is, this sequence gives a Cassini Identity of

- $12 * 13 - 1^2 = 156 - 1 = 155$

And this value of "155" = "5" * "31".

In the latter case again, we have a situation where the Rodin force graph resulting from multipliers "12" and "13" produces an algebraic progression. If we alter the first multiplier from "12" to "2" (or "-2", the non-spin corrected value of "29" [= L(7)]) and make a corresponding adaption to the multiplier "13" to obtain "16" [=L(3)^2] (spin corrected "-15"), we then get the Rodin type graph sequence:

Multiplier M1	M1 Spin Corrected	n	Multiplier M2	M2 Spin Corrected
15	15	1	1	1
8	8	2	29	-2
27	-4	3	4	4
2	2	4	23	-8
30	-1	5	16	-15
16	-15	6	30	-1
23	-8	7	2	2
4	4	8	27	-4
29	-2	9	8	8
1	1	10	15	15

Rodin-type graph sequence with multipliers "29" (also "-2") and "15" (also "-16").

An Aether expressing Fibonacci and Lucas Sequences joined at the hip: a thought experiment

Multiplier M1	M1 Spin Corrected	n	Multiplier M2	M2 Spin Corrected
1	1	1	12	12
13	13	2	20	-11
14	14	3	23	-8
27	-4	4	28	-3
10	10	5	26	-5
6	6	6	2	2
16	-15	7	24	-7
22	-9	8	9	9
7	7	9	15	15
29	-2	10	25	-6
5	5	11	21	-10
3	3	12	4	4
8	8	13	17	-14
11	11	14	18	-13
19	-12	15	30	-1
30	-1	16	19	-12
18	-13	17	11	11
17	-14	18	8	8
4	4	19	3	3
21	-10	20	5	5
25	-6	21	29	-2
15	15	22	7	7
9	9	23	22	-9
24	-7	24	16	-15
2	2	25	6	6
26	-5	26	10	10
28	-3	27	27	-4
23	-8	28	14	14
20	-11	29	13	13
12	12	30	1	1

Rodin-type graph sequence for Modulo-31 with multipliers "12" and "13".

An Aether expressing Fibonacci and Lucas Sequences joined at the hip: a thought experiment

Tables for MNS-L(4)^2 (Fibonacci Sequence)

Standard				n	Spin Corrected				
F(x)		F(-x)	F(x) * F(-x)	n	F(x)		F(-x)	F(x) * F(-x)	
	0		0	0		0		0	
1		1	1	1	1		1	1	
1		-1	-1	2	1		-1	-1	
2		2	4	3	2		2	4	
3		-3	-9	4	3		-3	-9	
5		5	25	5	5		5	-24	
8		-8	34	6	8		-8	-15	
13		13	22	7	13		13	22	
21		-21	0	8	21		-21	0	I
34		34	29	9	-15		-15	-20	
6		-6	-36	10	6		-6	13	
40		40	32	11	-9		-9	-17	
46		-46	40	12	-3		3	-9	
37		37	46	13	-12		-12	-3	
34		-34	20	14	-15		15	20	
22		22	43	15	22		22	-6	
7		-7	0	16	7		-7	0	I
29		29	8	17	-20		-20	8	
36		-36	27	18	-13		13	-22	
16		16	11	19	16		16	11	
3		-3	-9	20	3		-3	-9	
19		19	18	21	19		19	18	
22		-22	6	22	22		-22	6	
41		41	15	23	-8		-8	15	
14		-14	0	24	14		-14	0	I

The Fibonacci sequence with the MNS-49 circuit requires 112 iterations before the cycle begins again. For efficiency, only the first 24 are shown here.

273

An Aether expressing Fibonacci and Lucas Sequences joined at the hip: a thought experiment

Fibonacci word cycle MNS-49 circuit.

(using "0" and roots of "0" to punctuate words)

Carrier States				Supra-MNS states		
Original Sum		Corrected Sum	Word Index n	Sum (x+)		Sum (x-)
	0				0	
-22		-22	1	33		9
-22		-22	2	23		47
27		-22	3	19		44
-71		-22	4	37		12
27		-22	5	5		30
-22		-22	6	2		26
-22		-22	7	40		16
-22		-22	8	16		40
-22		-22	9	26		2
27		-22	10	30		5
-71		-22	11	12		37
27		-22	12	44		19
-22		-22	13	47		23
-22		-22	14	9		33

An Aether expressing Fibonacci and Lucas Sequences joined at the hip: a thought experiment

Fibonacci word cycle MNS-49 circuit.

(using "0" and roots of "0" to punctuate words)

Standard Supra-MNS states				Word Index n	Spin Corrected		
Words	Sum (x+)		Sum (x-)		Sum (x+)		Sum (x-)
		0				0	
Fr -Sum 1	33		9	1	-16		9
Fr -Sum 2	23		47	2	23		-2
Fr -Sum 3	19		44	3	19		-5
Fr -Sum 4	37		12	4	-12		12
Fr -Sum 5	5		30	5	5		-19
Fr -Sum 6	2		26	6	2		-23
Fr -Sum 7	40		16	7	-9		16
Fr -Sum 8	16		40	8	16		-9
Fr -Sum 9	26		2	9	-23		2
Fr -Sum 10	30		5	10	-19		5
Fr -Sum 11	12		37	11	12		-12
Fr -Sum 12	44		19	12	-5		19
Fr -Sum 13	47		23	13	-2		23
Fr -Sum 14	9		33	14	9		-16

An Aether expressing Fibonacci and Lucas Sequences joined at the hip: a thought experiment

Summary of Fibonacci Carrier and supra-MNS states for the MNS-L(4)^2 circuit

The „carrier-state" of MNS-49 = "-22" and has a complementary value of "27".

When filtered, this complementary values of "27" carrier state gives rise to values of:

- "-4[or 5]" [= -22] and "0" [= 27] in MNS-9 circuits; and

- "3" [= -22] and "2" [= 27] in MNS-25 circuits.

Depending upon your point of view, these carrier values of "-4 [or 5]" and "0" in MNS-9:

- counter resonates with the general "regeneration combination" of "0" and "4" states; and

- resonates with the multiplier "5" used to produce the reverse direction Supra-MNS states of MNS-9 circuit.

The multipliers of the Supra-MNS state for MNS-49 include "20" and "-22". These have respective complementary multipliers "-29" and "27".

An Aether expressing Fibonacci and Lucas Sequences joined at the hip: a thought experiment

Tables for MNS-L(4)^2 (Lucas Sequence)

Standard				n	Spin Corrected				
L(r)		L(-r)	L(r) * L(-r)	n	L(r)		L(-r)	L(r) * L(-r)	
	2		4			2		4	
1		-1	-1	1	1		-1	-1	
3		3	9	2	3		3	9	
4		-4	-16	3	4		-4	-16	*
7		7	0	4	7		7	0	I
11		-11	26	5	11		-11	-23	
18		18	30	6	18		18	-19	
29		-29	41	7	-20		20	-8	
47		47	4	8	-2		-2	4	*
27		-27	6	9	-22		22	6	
25		25	37	10	-24		-24	-12	
3		-3	-9	11	3		-3	-9	
28		28	0	12	-21		-21	0	I
31		-31	19	13	-18		18	19	
10		10	2	14	10		10	2	
41		-41	34	15	-8		8	-15	
2		2	4	16	2		2	4	*
43		-43	13	17	-6		6	13	
45		45	16	18	-4		-4	16	
39		-39	47	19	-10		10	-2	
35		35	0	20	-14		-14	0	I
25		-25	12	21	-24		24	12	
11		11	23	22	11		11	23	
36		-36	27	23	-13		13	-22	
47		47	4	24	-2		-2	4	*
34		-34	20	25	-15		15	20	
32		32	44	26	-17		-17	-5	
17		-17	5	27	17		-17	5	
0		0	0	28	0		0	0	I I

An Aether expressing Fibonacci and Lucas Sequences joined at the hip: a thought experiment

The Lucas sequence with the MNS-49 circuit requires 112 iterations before the cycle begins again. For efficiency, only the first 28 are shown here.

Lucas word cycle MNS-49 circuit.

(using "4" to punctuate words)

Carrier states				Supra-MNS states		
Original Sum		Corrected Sum	Word Index n	Sum (r+)		Sum (r-)
	4				4	
-58		-9	1	24		32
-9		-9	2	18		10
40		-9	3	38		46
40		-9	4	4		45
40		-9	5	3		11
-9		-9	6	39		31
-58		-9	7	17		25
-58		-9	8	25		17
-9		-9	9	31		39
40		-9	10	11		3
40		-9	11	45		4
40		-9	12	46		38
-9		-9	13	10		18
-58		-9	14	32		24

An Aether expressing Fibonacci and Lucas Sequences
joined at the hip: a thought experiment

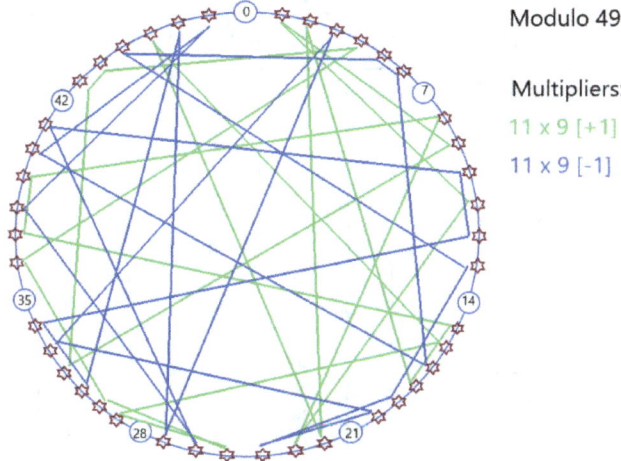

Modulo 49

Multipliers:
11 x 9 [+1]
11 x 9 [-1]

**Rodin-type force graph with complementary multipliers
"11" * "9" with (1) start-point "1" and with (2) start point
"-1"**

An Aether expressing Fibonacci and Lucas Sequences joined at the hip: a thought experiment

Lucas word cycle MNS-49 circuit.

(using "4" to punctuate words)

Standard Supra-MNS states				Word Index n	Spin Corrected		
Words	Sum (r+)		Sum (r-)		Sum (r+)		Sum (r-)
		4				4	
Fr -Sum 1	24		32	1	24		-17
Fr -Sum 2	18		10	2	18		10
Fr -Sum 3	38		46	3	-11		-3
Fr -Sum 4	4		45	4	4		-4
Fr -Sum 5	3		11	5	3		11
Fr -Sum 6	39		31	6	-10		-18
Fr -Sum 7	17		25	7	17		-24
Fr -Sum 8	25		17	8	-24		17
Fr -Sum 9	31		39	9	-18		-10
Fr -Sum 10	11		3	10	11		3
Fr -Sum 11	45		4	11	-4		4
Fr -Sum 12	46		38	12	-3		-11
Fr -Sum 13	10		18	13	10		18
Fr -Sum 14	32		24	14	-17		24

An Aether expressing Fibonacci and Lucas Sequences
joined at the hip: a thought experiment

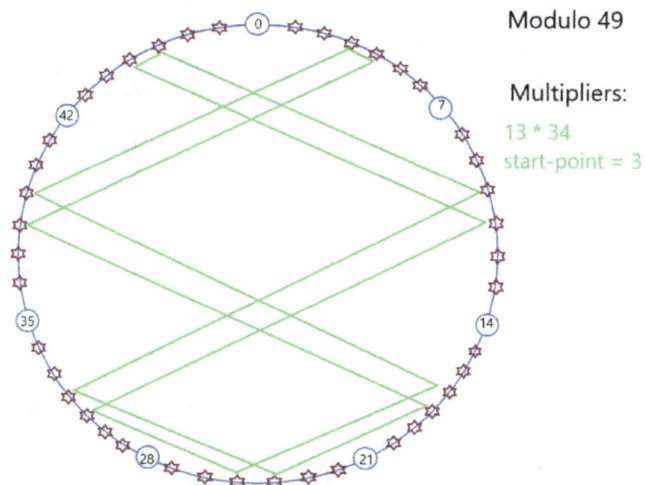

Modulo 49

Multipliers:

13 * 34
start-point = 3

A Rodin-type force graph with use of complementary
multipliers "13" [= F(7)] and "34" [= F(9)]. Complementary
multipliers F(7) and F(9) are also associated with the
Cassini Identifier for an MNS-F(8)² circuit.

An Aether expressing Fibonacci and Lucas Sequences joined at the hip: a thought experiment

Lucas word cycle MNS-49 circuit.

(using "0" and roots of "0" to punctuate words)

Carrier states				Supra-MNS states		
Original Sum		Corrected Sum	Word Index n	Sum (r+)		Sum (r-)
	4				4	
-61		-12	1	8		8
-61		-12	2	13		20
37		-12	3	15		1
37		-12	4	6		27
37		-12	5	22		43
37		-12	6	48		34
-61		-12	7	29		36
-12		-12	8	41		41
-61		-12	9	36		29
37		-12	10	34		48
37		-12	11	43		22
37		-12	12	27		6
37		-12	13	1		15
-61		-12	14	20		13
-12		-12	15	8		8

An Aether expressing Fibonacci and Lucas Sequences
joined at the hip: a thought experiment

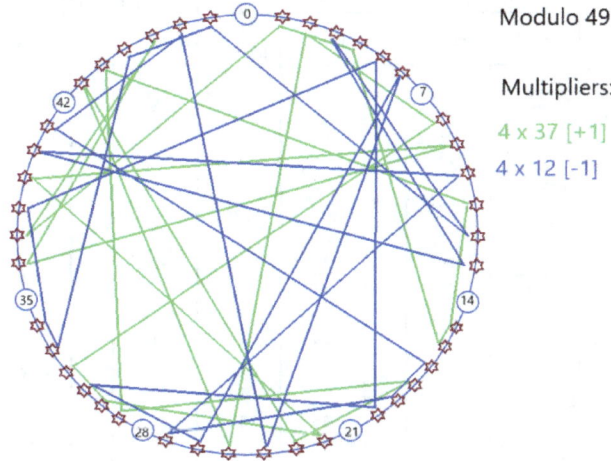

Modulo 49

Multipliers:

4 x 37 [+1]
4 x 12 [-1]

**Rodin-type force graph for MNS-L(4)² with
complementary multipliers "37" [= -12] and "4".**

An Aether expressing Fibonacci and Lucas Sequences joined at the hip: a thought experiment

Lucas word cycle MNS-49 circuit.

(using "0" and roots of "0" to punctuate words)

Standard Supra-MNS states				Word Index n	Spin Corrected		
Words	Sum (r+)		Sum (r-)		Sum (r+)		Sum (r-)
		4				4	
Fr -Sum 1	8		8	1	8		8
Fr -Sum 2	13		20	2	13		20
Fr -Sum 3	15		1	3	15		1
Fr -Sum 4	6		27	4	6		-22
Fr -Sum 5	22		43	5	22		-6
Fr -Sum 6	48		34	6	-1		-15
Fr -Sum 7	29		36	7	-20		-13
Fr -Sum 8	41		41	8	-8		-8
Fr -Sum 9	36		29	9	-13		-20
Fr -Sum 10	34		48	10	-15		-1
Fr -Sum 11	43		22	11	-6		22
Fr -Sum 12	27		6	12	-22		6
Fr -Sum 13	1		15	13	1		15
Fr -Sum 14	20		13	14	20		13
Fr -Sum 15	8		8	15	8		8

Being demarcated by "0" which is out of phase with the demarcation of "4", causes each Aether-state word to form out 180-degrees of phase with / overlap the corresponding words formed of "0"s in sequence tables of the Fibonacci sequence. Thus, the Aether-state "8" appears to be on both time strands simultaneously.

An Aether expressing Fibonacci and Lucas Sequences joined at the hip: a thought experiment

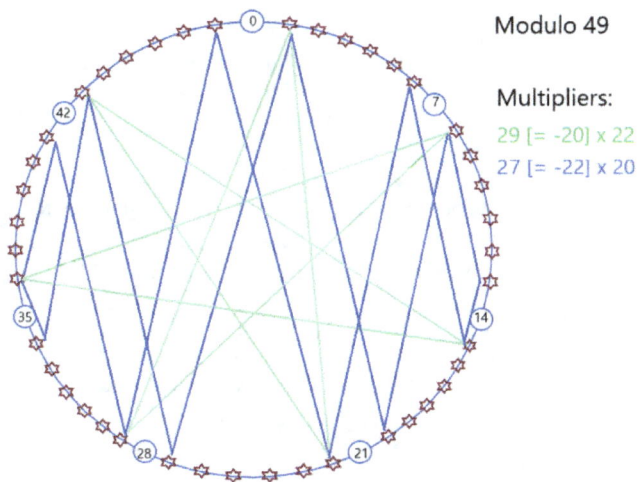

Modulo 49

Multipliers:

29 [= -20] x 22

27 [= -22] x 20

Rodin-type force graph for MNS-L(4)² with complementary multipliers "29" [= -20] and "22" (green) and complementary multipliers "27" [= -22] and "20"

An Aether expressing Fibonacci and Lucas Sequences joined at the hip: a thought experiment

Summary of Lucas Carrier and supra-MNS states for the MNS-L(4)^2 circuit

- the "0" and "4" demarcated "carrier"-states of the Lucas sequence have respective values of
    - "-9" [= 40] and
    - "-12" [= 37]

    i.e., rather than "-22" [= 27] of the Fibonacci Sequence, when filtered then through the MNS-9 circuit, the former Lucas "carrier"-state has a value of "0" and the latter has a value of "-3" i.e., root-"0";

- the "0"- demarcated "carrier"-state of "-9" for MNS-49 circuit also resonates directly with the corresponding "carrier"-state of MNS-25 when demarcated by "4";
- The "0"- demarcated "carrier"- state of "-9" also simultaneously filters in MNS-9 for a "regeneration combination" of "0" and "4";
- The Supra-MNS word cycle of the "0"-demarcated Lucas cycle is out of phase with, but repeats, the Supra-MNS word cycle of the "0"-demarcated Fibonacci cycle which uses the multipliers of "-22" [= 27] and "20" [-29 = L(-7)], to thereby produce a virtual MNS-L(6)^2 type centre. That is "-22" = "-2" * "11" = "-2" * L(5). Thus, the product of "2" * [L(5) * L(-7)] is one part of the Cassini Identity involving L(6)^2 minus the [L(5) * L(-7)] product;

An Aether expressing Fibonacci and Lucas Sequences joined at the hip: a thought experiment

- The Supra-MNS word cycle of the "4"-demarcated Lucas cycle comprises word values of 13 [= F(7)] and 34 [= F(9)] as complementary multipliers to produce the Supra-MNS states in question. These values per se correspond to values of multipliers of a Cassini Identity inversion line of an MNS-F(8)^2 [MNS-21^2] [= 9 x 49] circuit; and
- The contra values of "13" and "34", used as multipliers in MNS-49, are "36" and "15" respectively. Filtered through MNS-9, these contra-values amount to "0"[= 36] and root-"0"[= 3 * 5] respectively.

An Aether expressing Fibonacci and Lucas Sequences joined at the hip: a thought experiment

Epilogue 11:

Tables of Fibonacci and Lucas values for MNS-F(7)^2 circuit

What follows is a list of each of the 364 iterations for the Fibonacci Sequence and the Lucas sequence. Each seventh iteration of both the Fibonacci and Lucas sequences is marked with a yellow, blue, or pink background. The pink background indicates a cycle of 91 iterations has been completed.

An Aether expressing Fibonacci and Lucas Sequences joined at the hip: a thought experiment

Standard				n	Spin Corrected				
F(x)		F(-x)	F(x) * F(-x)	n	F(x)		F(-x)	F(x) * F(-x)	
	0		0	0		0		0	
1		1	1	1	1		1	1	
1		-1	-1	2	1		-1	-1	
2		2	4	3	2		2	4	
3		-3	-9	4	3		-3	-9	
5		5	25	5	5		5	25	
8		-8	-64	6	8		-8	-64	
13		13	0	7	13		13	0	I
21		-21	66	8	21		-21	66	
34		34	142	9	34		34	-27	
55		-55	17	10	55		-55	17	
89		89	147	11	-80		-80	-22	
144		-144	51	12	-25		25	51	
64		64	40	13	64		64	40	
39		-39	0	14	39		-39	0	I
103		103	131	15	-66		-66	-38	
142		-142	116	16	-27		27	-53	
76		76	30	17	76		76	30	
49		-49	134	18	49		-49	-35	
125		125	77	19	-44		-44	77	
5		-5	-25	20	5		-5	-25	
130		130	0	21	-39		-39	0	I
135		-135	27	22	-34		34	27	
96		96	90	23	-73		-73	-79	
62		-62	43	24	62		-62	43	
158		158	121	25	-11		-11	-48	
51		-51	103	26	51		-51	-66	
40		40	79	27	40		40	79	
91		-91	0	28	-78		78	0	I
131		131	92	29	-38		-38	-77	
53		-53	64	30	53		-53	64	
15		15	56	31	15		15	56	
68		-68	108	32	68		-68	-61	
83		83	129	33	83		83	-40	
151		-151	14	34	-18		18	14	
65		65	0	35	65		65	0	I
47		-47	157	36	47		-47	-12	
112		112	38	37	-57		-57	38	
159		-159	69	38	-10		10	69	
102		102	95	39	-67		-67	-74	
92		-92	155	40	-77		77	-14	
25		25	118	41	25		25	-51	
117		-117	0	42	-52		52	0	I

An Aether expressing Fibonacci and Lucas Sequences joined at the hip: a thought experiment

Standard				n	Spin Corrected			
L(r)		L(-r)	L(r) * L(-r)		L(r)		L(-r)	L(r) * L(-r)
	2		4			2		4
1		-1	-1	1	1		-1	-1
3		3	9	2	3		3	9
4		-4	-16	3	4		-4	-16
7		7	49	4	7		7	49
11		-11	-121	5	11		-11	48
18		18	155	6	18		18	-14
29		-29	4	7	29		-29	4
47		47	12	8	47		47	12
76		-76	139	9	76		-76	-30
123		123	88	10	-46		-46	-81
30		-30	114	11	30		-30	-55
153		153	87	12	-16		-16	-82
14		-14	142	13	14		-14	-27
167		167	4	14	-2		-2	4
12		-12	-144	15	12		-12	25
10		10	100	16	10		10	-69
22		-22	23	17	22		-22	23
32		32	10	18	32		32	10
54		-54	126	19	54		-54	-43
86		86	129	20	-83		-83	-40
140		-140	4	21	-29		29	4
57		57	38	22	57		57	38
28		-28	61	23	28		-28	61
85		85	127	24	-84		-84	-42
113		-113	75	25	-56		56	75
29		29	165	26	29		29	-4
142		-142	116	27	-27		27	-53
2		2	4	28	2		2	4
144		-144	51	29	-25		25	51
146		146	22	30	-23		-23	22
121		-121	62	31	-48		48	62
98		98	140	32	-71		-71	-29
50		-50	35	33	50		-50	35
148		148	103	34	-21		-21	-66
29		-29	4	35	29		-29	4
8		8	64	36	8		8	64
37		-37	152	37	37		-37	-17
45		45	166	38	45		45	-3
82		-82	36	39	82		-82	36
127		127	74	40	-42		-42	74
40		-40	90	41	40		-40	-79
167		167	4	42	-2		-2	4

An Aether expressing Fibonacci and Lucas Sequences joined at the hip: a thought experiment

142	142	53	**43**	-27	-27	53
90	-90	12	**44**	-79	79	12
63	63	82	**45**	63	63	82
153	-153	82	**46**	-16	16	82
47	47	12	**47**	47	47	12
31	-31	53	**48**	31	-31	53
78	78	0	**49**	78	78	0
109	-109	118	**50**	-60	60	-51
18	18	155	**51**	18	18	-14
127	-127	95	**52**	-42	42	-74
145	145	69	**53**	-24	-24	69
103	-103	38	**54**	-66	66	38
79	79	157	**55**	79	79	-12
13	-13	0	**56**	13	-13	0
92	92	14	**57**	-77	-77	14
105	-105	129	**58**	-64	64	-40
28	28	108	**59**	28	28	-61
133	-133	56	**60**	-36	36	56
161	161	64	**61**	-8	-8	64
125	-125	92	**62**	-44	44	-77
117	117	0	**63**	-52	-52	0
73	-73	79	**64**	73	-73	79
21	21	103	**65**	21	21	-66
94	-94	121	**66**	-75	75	-48
115	115	43	**67**	-54	-54	43
40	-40	90	**68**	40	-40	-79
155	155	27	**69**	-14	-14	27
26	-26	0	**70**	26	-26	0
12	12	144	**71**	12	12	-25
38	-38	77	**72**	38	-38	77
50	50	134	**73**	50	50	-35
88	-88	30	**74**	-81	81	30
138	138	116	**75**	-31	-31	-53
57	-57	131	**76**	57	-57	-38
26	26	0	**77**	26	26	0
83	-83	40	**78**	83	-83	40
109	109	51	**79**	-60	-60	51
23	-23	147	**80**	23	-23	-22
132	132	17	**81**	-37	-37	17
155	-155	142	**82**	-14	14	-27
118	118	66	**83**	-51	-51	66
104	-104	0	**84**	-65	65	0
53	53	105	**85**	53	53	-64
157	-157	25	**86**	-12	12	25
41	41	160	**87**	41	41	-9
29	-29	4	**88**	29	-29	4
70	70	168	**89**	70	70	-1
99	-99	1	**90**	-70	70	1
0	0	0	**91**	0	0	0

An Aether expressing Fibonacci and Lucas Sequences joined at the hip: a thought experiment

38	-38	77	43	38	-38	77	
36	36	113	44	36	36	-56	
74	-74	101	45	74	-74	-68	
110	110	101	46	-59	-59	-68	
15	-15	113	47	15	-15	-56	
125	125	77	48	-44	-44	77	
140	-140	4	49	-29	29	4	*
96	96	90	50	-73	-73	-79	
67	-67	74	51	67	-67	74	
163	163	36	52	-6	-6	36	
61	-61	166	53	61	-61	-3	
55	55	152	54	55	55	-17	
116	-116	64	55	-53	53	64	
2	2	4	56	2	2	4	*
118	-118	103	57	-51	51	-66	
120	120	35	58	-49	-49	35	
69	-69	140	59	69	-69	-29	
20	20	62	60	20	20	62	
89	-89	22	61	-80	80	22	
109	109	51	62	-60	-60	51	
29	-29	4	63	29	-29	4	*
138	138	116	64	-31	-31	-53	
167	-167	165	65	-2	2	-4	
136	136	75	66	-33	-33	75	
134	-134	127	67	-35	35	-42	
101	101	61	68	-68	-68	61	
66	-66	38	69	66	-66	38	
167	167	4	70	-2	-2	4	*
64	-64	129	71	64	-64	-40	
62	62	126	72	62	62	-43	
126	-126	10	73	-43	43	10	
19	19	23	74	19	19	23	
145	-145	100	75	-24	24	-69	
164	164	25	76	-5	-5	25	
140	-140	4	77	-29	29	4	*
135	135	142	78	-34	-34	-27	
106	-106	87	79	-63	63	-82	
72	72	114	80	72	72	-55	
9	-9	-81	81	9	-9	-81	
81	81	139	82	81	81	-30	
90	-90	12	83	-79	79	12	
2	2	4	84	2	2	4	*
92	-92	155	85	-77	77	-14	
94	94	48	86	-75	-75	48	
17	-17	49	87	17	-17	49	
111	111	153	88	-58	-58	-16	
128	-128	9	89	-41	41	9	
70	70	168	90	70	70	-1	
29	-29	4	91	29	-29	4	*

An Aether expressing Fibonacci and Lucas Sequences joined at the hip: a thought experiment

Standard			n	Spin Corrected		
F(x)	F(-x)	F(x) * F(-x)		F(x)	F(-x)	F(x) * F(-x)
0		0	0	0		0
99	-99	1	92	-70	70	1
99	99	168	93	-70	-70	-1
29	-29	4	94	29	-29	4
128	128	160	95	-41	-41	-9
157	-157	25	96	-12	12	25
116	116	105	97	-53	-53	-64
104	-104	0	98	-65	65	0
51	51	66	99	51	51	66
155	-155	142	100	-14	14	-27
37	37	17	101	37	37	17
23	-23	147	102	23	-23	-22
60	60	51	103	60	60	51
83	-83	40	104	83	-83	40
143	143	0	105	-26	-26	0
57	-57	131	106	57	-57	-38
31	31	116	107	31	31	-53
88	-88	30	108	-81	81	30
119	119	134	109	-50	-50	-35
38	-38	77	110	38	-38	77
157	157	144	111	-12	-12	-25
26	-26	0	112	26	-26	0
14	14	27	113	14	14	27
40	-40	90	114	40	-40	-79
54	54	43	115	54	54	43
94	-94	121	116	-75	75	-48
148	148	103	117	-21	-21	-66
73	-73	79	118	73	-73	79
52	52	0	119	52	52	0
125	-125	92	120	-44	44	-77
8	8	64	121	8	8	64
133	-133	56	122	-36	36	56
141	141	108	123	-28	-28	-61
105	-105	129	124	-64	64	-40
77	77	14	125	77	77	14
13	-13	0	126	13	-13	0
90	90	157	127	-79	-79	-12
103	-103	38	128	-66	66	38
24	24	69	129	24	24	69
127	-127	95	130	-42	42	-74
151	151	155	131	-18	-18	-14
109	-109	118	132	-60	60	-51
91	91	0	133	-78	-78	0

An Aether expressing Fibonacci and Lucas Sequences joined at the hip: a thought experiment

Standard				n	Spin Corrected			
L(r)		L(-r)	L(r) * L(-r)		L(r)	L(-r)	L(r) * L(-r)	
	2		4			2	4	
99		99	168	92	-70	-70	-1	
128		-128	9	93	-41	41	9	
58		58	153	94	58	58	-16	
17		-17	49	95	17	-17	49	
75		75	48	96	75	75	48	
92		-92	155	97	-77	77	-14	
167		167	4	98	-2	-2	4	*
90		-90	12	99	-79	79	12	
88		88	139	100	-81	-81	-30	
9		-9	-81	101	9	-9	-81	
97		97	114	102	-72	-72	-55	
106		-106	87	103	-63	63	-82	
34		34	142	104	34	34	-27	
140		-140	4	105	-29	29	4	*
5		5	25	106	5	5	25	
145		-145	100	107	-24	24	-69	
150		150	23	108	-19	-19	23	
126		-126	10	109	-43	43	10	
107		107	126	110	-62	-62	-43	
64		-64	129	111	64	-64	-40	
2		2	4	112	2	2	4	*
66		-66	38	113	66	-66	38	
68		68	61	114	68	68	61	
134		-134	127	115	-35	35	-42	
33		33	75	116	33	33	75	
167		-167	165	117	-2	2	-4	
31		31	116	118	31	31	-53	
29		-29	4	119	29	-29	4	*
60		60	51	120	60	60	51	
89		-89	22	121	-80	80	22	
149		149	62	122	-20	-20	62	
69		-69	140	123	69	-69	-29	
49		49	35	124	49	49	35	
118		-118	103	125	-51	51	-66	
167		167	4	126	-2	-2	4	*
116		-116	64	127	-53	53	64	
114		114	152	128	-55	-55	-17	
61		-61	166	129	61	-61	-3	
6		6	36	130	6	6	36	
67		-67	74	131	67	-67	74	
73		73	90	132	73	73	-79	
140		-140	4	133	-29	29	4	*

An Aether expressing Fibonacci and Lucas Sequences joined at the hip: a thought experiment

31	-31	53	134	31	-31	53	
122	122	12	135	-47	-47	12	
153	-153	82	136	-16	16	82	
106	106	82	137	-63	-63	82	
90	-90	12	138	-79	79	12	
27	27	53	139	27	27	53	
117	-117	0	140	-52	52	0	I
144	144	118	141	-25	-25	-51	
92	-92	155	142	-77	77	-14	
67	67	95	143	67	67	-74	
159	-159	69	144	-10	10	69	
57	57	38	145	57	57	38	
47	-47	157	146	47	-47	-12	
104	104	0	147	-65	-65	0	I
151	-151	14	148	-18	18	14	
86	86	129	149	-83	-83	-40	
68	-68	108	150	68	-68	-61	
154	154	56	151	-15	-15	56	
53	-53	64	152	53	-53	64	
38	38	92	153	38	38	-77	
91	-91	0	154	-78	78	0	I
129	129	79	155	-40	-40	79	
51	-51	103	156	51	-51	-66	
11	11	121	157	11	11	-48	
62	-62	43	158	62	-62	43	
73	73	90	159	73	73	-79	
135	-135	27	160	-34	34	27	
39	39	0	161	39	39	0	I
5	-5	-25	162	5	-5	-25	
44	44	77	163	44	44	77	
49	-49	134	164	49	-49	-35	
93	93	30	165	-76	-76	30	
142	-142	116	166	-27	27	-53	
66	66	131	167	66	66	-38	
39	-39	0	168	39	-39	0	I
105	105	40	169	-64	-64	40	
144	-144	51	170	-25	25	51	
80	80	147	171	80	80	-22	
55	-55	17	172	55	-55	17	
135	135	142	173	-34	-34	-27	
21	-21	66	174	21	-21	66	
156	156	0	175	-13	-13	0	I
8	-8	-64	176	8	-8	-64	
164	164	25	177	-5	-5	25	
3	-3	-9	178	3	-3	-9	
167	167	4	179	-2	-2	4	
1	-1	-1	180	1	-1	-1	
168	168	1	181	-1	-1	1	
0	0	0	182	0	0	0	I I

An Aether expressing Fibonacci and Lucas Sequences joined at the hip: a thought experiment

44	44	77	**134**	44	44	77	
15	-15	113	**135**	15	-15	-56	
59	59	101	**136**	59	59	-68	
74	-74	101	**137**	74	-74	-68	
133	133	113	**138**	-36	-36	-56	
38	-38	77	**139**	38	-38	77	
2	2	4	140	2	2	4	*
40	-40	90	**141**	40	-40	-79	
42	42	74	**142**	42	42	74	
82	-82	36	**143**	82	-82	36	
124	124	166	**144**	-45	-45	-3	
37	-37	152	**145**	37	-37	-17	
161	161	64	**146**	-8	-8	64	
29	-29	4	147	29	-29	4	*
21	21	103	**148**	21	21	-66	
50	-50	35	**149**	50	-50	35	
71	71	140	**150**	71	71	-29	
121	-121	62	**151**	-48	48	62	
23	23	22	**152**	23	23	22	
144	-144	51	**153**	-25	25	51	
167	167	4	154	-2	-2	4	*
142	-142	116	**155**	-27	27	-53	
140	140	165	**156**	-29	-29	-4	
113	-113	75	**157**	-56	56	75	
84	84	127	**158**	84	84	-42	
28	-28	61	**159**	28	-28	61	
112	112	38	**160**	-57	-57	38	
140	-140	4	161	-29	29	4	*
83	83	129	**162**	83	83	-40	
54	-54	126	**163**	54	-54	-43	
137	137	10	**164**	-32	-32	10	
22	-22	23	**165**	22	-22	23	
159	159	100	**166**	-10	-10	-69	
12	-12	-144	**167**	12	-12	25	
2	2	4	168	2	2	4	*
14	-14	142	**169**	14	-14	-27	
16	16	87	**170**	16	16	-82	
30	-30	114	**171**	30	-30	-55	
46	46	88	**172**	46	46	-81	
76	-76	139	**173**	76	-76	-30	
122	122	12	**174**	-47	-47	12	
29	-29	4	175	29	-29	4	*
151	151	155	**176**	-18	-18	-14	
11	-11	-121	**177**	11	-11	48	
162	162	49	**178**	-7	-7	49	
4	-4	-16	179	4	-4	-16	*
166	166	9	**180**	-3	-3	9	
1	-1	-1	**181**	1	-1	-1	
167	167	4	182	-2	-2	4	*

An Aether expressing Fibonacci and Lucas Sequences joined at the hip: a thought experiment

Standard				n	Spin Corrected		
F(x)		F(-x)	F(x) * F(-x)	n	F(x)	F(-x)	F(x) * F(-x)
	0		0	0	0		0
168		168	1	183	-1	-1	1
168		-168	168	184	-1	1	-1
167		167	4	185	-2	-2	4
166		-166	160	186	-3	3	-9
164		164	25	187	-5	-5	25
161		-161	105	188	-8	8	-64
156		156	0	189	-13	-13	0	I
148		-148	66	190	-21	21	66
135		135	142	191	-34	-34	-27
114		-114	17	192	-55	55	17
80		80	147	193	80	80	-22
25		-25	51	194	25	-25	51
105		105	40	195	-64	-64	40
130		-130	0	196	-39	39	0	I
66		66	131	197	66	66	-38
27		-27	116	198	27	-27	-53
93		93	30	199	-76	-76	30
120		-120	134	200	-49	49	-35
44		44	77	201	44	44	77
164		-164	144	202	-5	5	-25
39		39	0	203	39	39	0	I
34		-34	27	204	34	-34	27
73		73	90	205	73	73	-79
107		-107	43	206	-62	62	43
11		11	121	207	11	11	-48
118		-118	103	208	-51	51	-66
129		129	79	209	-40	-40	79
78		-78	0	210	78	-78	0	I
38		38	92	211	38	38	-77
116		-116	64	212	-53	53	64
154		154	56	213	-15	-15	56
101		-101	108	214	-68	68	-61
86		86	129	215	-83	-83	-40
18		-18	14	216	18	-18	14
104		104	0	217	-65	-65	0	I
122		-122	157	218	-47	47	-12
57		57	38	219	57	57	38
10		-10	-100	220	10	-10	69
67		67	95	221	67	67	-74
77		-77	155	222	77	-77	-14
144		144	118	223	-25	-25	-51
52		-52	0	224	52	-52	0	I

An Aether expressing Fibonacci and Lucas Sequences joined at the hip: a thought experiment

Standard			n	Spin Corrected			
L(r)	L(-r)	L(r) * L(-r)		L(r)	L(-r)	L(r) * L(-r)	
2		4			2	4	
168	-168	168	183	-1	1	-1	
166	166	9	184	-3	-3	9	
165	-165	153	185	-4	4	-16	
162	162	49	186	-7	-7	49	
158	-158	48	187	-11	11	48	
151	151	155	188	-18	-18	-14	
140	-140	4	189	-29	29	4	*
122	122	12	190	-47	-47	12	
93	-93	139	191	-76	76	-30	
46	46	88	192	46	46	-81	
139	-139	114	193	-30	30	-55	
16	16	87	194	16	16	-82	
155	-155	142	195	-14	14	-27	
2	2	4	196	2	2	4	*
157	-157	25	197	-12	12	25	
159	159	100	198	-10	-10	-69	
147	-147	23	199	-22	22	23	
137	137	10	200	-32	-32	10	
115	-115	126	201	-54	54	-43	
83	83	129	202	83	83	-40	
29	-29	4	203	29	-29	4	*
112	112	38	204	-57	-57	38	
141	-141	61	205	-28	28	61	
84	84	127	206	84	84	-42	
56	-56	75	207	56	-56	75	
140	140	165	208	-29	-29	-4	
27	-27	116	209	27	-27	-53	
167	167	4	210	-2	-2	4	*
25	-25	51	211	25	-25	51	
23	23	22	212	23	23	22	
48	-48	62	213	48	-48	62	
71	71	140	214	71	71	-29	
119	-119	35	215	-50	50	35	
21	21	103	216	21	21	-66	
140	-140	4	217	-29	29	4	*
161	161	64	218	-8	-8	64	
132	-132	152	219	-37	37	-17	
124	124	166	220	-45	-45	-3	
87	-87	36	221	-82	82	36	
42	42	74	222	42	42	74	
129	-129	90	223	-40	40	-79	
2	2	4	224	2	2	4	*

An Aether expressing Fibonacci and Lucas Sequences joined at the hip: a thought experiment

27	27	53	**225**	27	27	53	
79	-79	12	**226**	79	-79	12	
106	106	82	**227**	-63	-63	82	
16	-16	82	**228**	16	-16	82	
122	122	12	**229**	-47	-47	12	
138	-138	53	**230**	-31	31	53	
91	91	0	**231**	-78	-78	0	I
60	-60	118	**232**	60	-60	-51	
151	151	155	**233**	-18	-18	-14	
42	-42	95	**234**	42	-42	-74	
24	24	69	**235**	24	24	69	
66	-66	38	**236**	66	-66	38	
90	90	157	**237**	-79	-79	-12	
156	-156	0	**238**	-13	13	0	I
77	77	14	**239**	77	77	14	
64	-64	129	**240**	64	-64	-40	
141	141	108	**241**	-28	-28	-61	
36	-36	56	**242**	36	-36	56	
8	8	64	**243**	8	8	64	
44	-44	92	**244**	44	-44	-77	
52	52	0	**245**	52	52	0	I
96	-96	79	**246**	-73	73	79	
148	148	103	**247**	-21	-21	-66	
75	-75	121	**248**	75	-75	-48	
54	54	43	**249**	54	54	43	
129	-129	90	**250**	-40	40	-79	
14	14	27	**251**	14	14	27	
143	-143	0	**252**	-26	26	0	I
157	157	144	**253**	-12	-12	-25	
131	-131	77	**254**	-38	38	77	
119	119	134	**255**	-50	-50	-35	
81	-81	30	**256**	81	-81	30	
31	31	116	**257**	31	31	-53	
112	-112	131	**258**	-57	57	-38	
143	143	0	**259**	-26	-26	0	I
86	-86	40	**260**	-83	83	40	
60	60	51	**261**	60	60	51	
146	-146	147	**262**	-23	23	-22	
37	37	17	**263**	37	37	17	
14	-14	142	**264**	14	-14	-27	
51	51	66	**265**	51	51	66	
65	-65	0	**266**	65	-65	0	I
116	116	105	**267**	-53	-53	-64	
12	-12	-144	**268**	12	-12	25	
128	128	160	**269**	-41	-41	-9	
140	-140	4	**270**	-29	29	4	
99	99	168	**271**	-70	-70	-1	
70	-70	1	**272**	70	-70	1	
0	0	0	273	0	0	0	I I

An Aether expressing Fibonacci and Lucas Sequences joined at the hip: a thought experiment

131	-131	77	**225**	-38	38	77	
133	133	113	**226**	-36	-36	-56	
95	-95	101	**227**	-74	74	-68	
59	59	101	**228**	59	59	-68	
154	-154	113	**229**	-15	15	-56	
44	44	77	**230**	44	44	77	
29	-29	4	**231**	29	-29	4	*
73	73	90	**232**	73	73	-79	
102	-102	74	**233**	-67	67	74	
6	6	36	**234**	6	6	36	
108	-108	166	**235**	-61	61	-3	
114	114	152	**236**	-55	-55	-17	
53	-53	64	**237**	53	-53	64	
167	167	4	**238**	-2	-2	4	*
51	-51	103	**239**	51	-51	-66	
49	49	35	**240**	49	49	35	
100	-100	140	**241**	-69	69	-29	
149	149	62	**242**	-20	-20	62	
80	-80	22	**243**	80	-80	22	
60	60	51	**244**	60	60	51	
140	-140	4	**245**	-29	29	4	*
31	31	116	**246**	31	31	-53	
2	-2	-4	**247**	2	-2	-4	
33	33	75	**248**	33	33	75	
35	-35	127	**249**	35	-35	-42	
68	68	61	**250**	68	68	61	
103	-103	38	**251**	-66	66	38	
2	2	4	**252**	2	2	4	*
105	-105	129	**253**	-64	64	-40	
107	107	126	**254**	-62	-62	-43	
43	-43	10	**255**	43	-43	10	
150	150	23	**256**	-19	-19	23	
24	-24	100	**257**	24	-24	-69	
5	5	25	**258**	5	5	25	
29	-29	4	**259**	29	-29	4	*
34	34	142	**260**	34	34	-27	
63	-63	87	**261**	63	-63	-82	
97	97	114	**262**	-72	-72	-55	
160	-160	88	**263**	-9	9	-81	
88	88	139	**264**	-81	-81	-30	
79	-79	12	**265**	79	-79	12	
167	167	4	**266**	-2	-2	4	*
77	-77	155	**267**	77	-77	-14	
75	75	48	**268**	75	75	48	
152	-152	49	**269**	-17	17	49	
58	58	153	**270**	58	58	-16	
41	-41	9	**271**	41	-41	9	
99	99	168	**272**	-70	-70	-1	
140	-140	4	**273**	-29	29	4	*

An Aether expressing Fibonacci and Lucas Sequences joined at the hip: a thought experiment

Standard				n	Spin Corrected		
F(x)		F(-x)	F(x) * F(-x)	n	F(x)	F(-x)	F(x) * F(-x)
	0		0	0	0		0
70		-70	1	274	70	-70	1
70		70	168	275	70	70	-1
140		-140	4	276	-29	29	4
41		41	160	277	41	41	-9
12		-12	-144	278	12	-12	25
53		53	105	279	53	53	-64
65		-65	0	280	65	-65	0	I
118		118	66	281	-51	-51	66
14		-14	142	282	14	-14	-27
132		132	17	283	-37	-37	17
146		-146	147	284	-23	23	-22
109		109	51	285	-60	-60	51
86		-86	40	286	-83	83	40
26		26	0	287	26	26	0	I
112		-112	131	288	-57	57	-38
138		138	116	289	-31	-31	-53
81		-81	30	290	81	-81	30
50		50	134	291	50	50	-35
131		-131	77	292	-38	38	77
12		12	144	293	12	12	-25
143		-143	0	294	-26	26	0	I
155		155	27	295	-14	-14	27
129		-129	90	296	-40	40	-79
115		115	43	297	-54	-54	43
75		-75	121	298	75	-75	-48
21		21	103	299	21	21	-66
96		-96	79	300	-73	73	79
117		117	0	301	-52	-52	0	I
44		-44	92	302	44	-44	-77
161		161	64	303	-8	-8	64
36		-36	56	304	36	-36	56
28		28	108	305	28	28	-61
64		-64	129	306	64	-64	-40
92		92	14	307	-77	-77	14
156		-156	0	308	-13	13	0	I
79		79	157	309	79	79	-12
66		-66	38	310	66	-66	38
145		145	69	311	-24	-24	69
42		-42	95	312	42	-42	-74
18		18	155	313	18	18	-14
60		-60	118	314	60	-60	-51
78		78	0	315	78	78	0	I

An Aether expressing Fibonacci and Lucas Sequences joined at the hip: a thought experiment

Standard			n	Spin Corrected			
L(r)	L(-r)	L(r) * L(-r)		L(r)	L(-r)	L(r) * L(-r)	
2		4		2		4	
70	70	168	274	70	70	-1	
41	-41	9	275	41	-41	9	
111	111	153	276	-58	-58	-16	
152	-152	49	277	-17	17	49	
94	94	48	278	-75	-75	48	
77	-77	155	279	77	-77	-14	
2	2	4	280	2	2	4	*
79	-79	12	281	79	-79	12	
81	81	139	282	81	81	-30	
160	-160	88	283	-9	9	-81	
72	72	114	284	72	72	-55	
63	-63	87	285	63	-63	-82	
135	135	142	286	-34	-34	-27	
29	-29	4	287	29	-29	4	*
164	164	25	288	-5	-5	25	
24	-24	100	289	24	-24	-69	
19	19	23	290	19	19	23	
43	-43	10	291	43	-43	10	
62	62	126	292	62	62	-43	
105	-105	129	293	-64	64	-40	
167	167	4	294	-2	-2	4	*
103	-103	38	295	-66	66	38	
101	101	61	296	-68	-68	61	
35	-35	127	297	35	-35	-42	
136	136	75	298	-33	-33	75	
2	-2	-4	299	2	-2	-4	
138	138	116	300	-31	-31	-53	
140	-140	4	301	-29	29	4	*
109	109	51	302	-60	-60	51	
80	-80	22	303	80	-80	22	
20	20	62	304	20	20	62	
100	-100	140	305	-69	69	-29	
120	120	35	306	-49	-49	35	
51	-51	103	307	51	-51	-66	
2	2	4	308	2	2	4	*
53	-53	64	309	53	-53	64	
55	55	152	310	55	55	-17	
108	-108	166	311	-61	61	-3	
163	163	36	312	-6	-6	36	
102	-102	74	313	-67	67	74	
96	96	90	314	-73	-73	-79	
29	-29	4	315	29	-29	4	*

An Aether expressing Fibonacci and Lucas Sequences joined at the hip: a thought experiment

138	-138	53	**316**	-31	31	53	
47	47	12	**317**	47	47	12	
16	-16	82	**318**	16	-16	82	
63	63	82	**319**	63	63	82	
79	-79	12	**320**	79	-79	12	
142	142	53	**321**	-27	-27	53	
52	-52	0	**322**	52	-52	0	I
25	25	118	**323**	25	25	-51	
77	-77	155	**324**	77	-77	-14	
102	102	95	**325**	-67	-67	-74	
10	-10	-100	**326**	10	-10	69	
112	112	38	**327**	-57	-57	38	
122	-122	157	**328**	-47	47	-12	
65	65	0	**329**	65	65	0	I
18	-18	14	**330**	18	-18	14	
83	83	129	**331**	83	83	-40	
101	-101	108	**332**	-68	68	-61	
15	15	56	**333**	15	15	56	
116	-116	64	**334**	-53	53	64	
131	131	92	**335**	-38	-38	-77	
78	-78	0	**336**	78	-78	0	I
40	40	79	**337**	40	40	79	
118	-118	103	**338**	-51	51	-66	
158	158	121	**339**	-11	-11	-48	
107	-107	43	**340**	-62	62	43	
96	96	90	**341**	-73	-73	-79	
34	-34	27	**342**	34	-34	27	
130	130	0	**343**	-39	-39	0	I
164	-164	144	**344**	-5	5	-25	
125	125	77	**345**	-44	-44	77	
120	-120	134	**346**	-49	49	-35	
76	76	30	**347**	76	76	30	
27	-27	116	**348**	27	-27	-53	
103	103	131	**349**	-66	-66	-38	
130	-130	0	**350**	-39	39	0	I
64	64	40	**351**	64	64	40	
25	-25	51	**352**	25	-25	51	
89	89	147	**353**	-80	-80	-22	
114	-114	17	**354**	-55	55	17	
34	34	142	**355**	34	34	-27	
148	-148	66	**356**	-21	21	66	
13	13	0	**357**	13	13	0	I
161	-161	105	**358**	-8	8	-64	
5	5	25	**359**	5	5	25	
166	-166	160	**360**	-3	3	-9	
2	2	4	**361**	2	2	4	
168	-168	168	**362**	-1	1	-1	
1	1	1	**363**	1	1	1	
0	0	0	364	0	0	0	I I

An Aether expressing Fibonacci and Lucas Sequences joined at the hip: a thought experiment

125	125	77	**316**	-44	-44	77	
154	-154	113	**317**	-15	15	-56	
110	110	101	**318**	-59	-59	-68	
95	-95	101	**319**	-74	74	-68	
36	36	113	**320**	36	36	-56	
131	-131	77	**321**	-38	38	77	
167	167	4	**322**	-2	-2	4	*
129	-129	90	**323**	-40	40	-79	
127	127	74	**324**	-42	-42	74	
87	-87	36	**325**	-82	82	36	
45	45	166	**326**	45	45	-3	
132	-132	152	**327**	-37	37	-17	
8	8	64	**328**	8	8	64	
140	-140	4	**329**	-29	29	4	*
148	148	103	**330**	-21	-21	-66	
119	-119	35	**331**	-50	50	35	
98	98	140	**332**	-71	-71	-29	
48	-48	62	**333**	48	-48	62	
146	146	22	**334**	-23	-23	22	
25	-25	51	**335**	25	-25	51	
2	2	4	**336**	2	2	4	*
27	-27	116	**337**	27	-27	-53	
29	29	165	**338**	29	29	-4	
56	-56	75	**339**	56	-56	75	
85	85	127	**340**	-84	-84	-42	
141	-141	61	**341**	-28	28	61	
57	57	38	**342**	57	57	38	
29	-29	4	**343**	29	-29	4	*
86	86	129	**344**	-83	-83	-40	
115	-115	126	**345**	-54	54	-43	
32	32	10	**346**	32	32	10	
147	-147	23	**347**	-22	22	23	
10	10	100	**348**	10	10	-69	
157	-157	25	**349**	-12	12	25	
167	167	4	**350**	-2	-2	4	*
155	-155	142	**351**	-14	14	-27	
153	153	87	**352**	-16	-16	-82	
139	-139	114	**353**	-30	30	-55	
123	123	88	**354**	-46	-46	-81	
93	-93	139	**355**	-76	76	-30	
47	47	12	**356**	47	47	12	
140	-140	4	**357**	-29	29	4	*
18	18	155	**358**	18	18	-14	
158	-158	48	**359**	-11	11	48	
7	7	49	**360**	7	7	49	
165	-165	153	**361**	-4	4	-16	
3	3	9	**362**	3	3	9	
168	-168	168	**363**	-1	1	-1	
2	2	4	364	2	2	4	*

An Aether expressing Fibonacci and Lucas Sequences joined at the hip: a thought experiment

An alternative view of what a proton comprises.

From what has been deduced through measurement, the mass of a proton is approximately 1,836 times the mass of an electron [23]. For this exercise I will take this number at face value and assume a proton in some way or manifestation comprises 1,836 electrons.

So, let's:

- first factorise the number 1,836 into some constituent factors; and
- then, based on Rodin-type force graphs, examine possible structures which could be created which requires 1,836 electrons.

1,836 can be factorised into 18 * 102.

102 can be factorised into 2 * 51.

51 can be factorised into 3 * 17.

Now factors 3 [= F(4)] * 17 [= 25 − F(6)] are Cassini Identity factors involving the MNS-F(5)^2 circuit.

We can however create a harmonic of this MNS-F(5)^2 circuit with 2 * 25 Aether-elements.

An Aether expressing Fibonacci and Lucas Sequences joined at the hip: a thought experiment

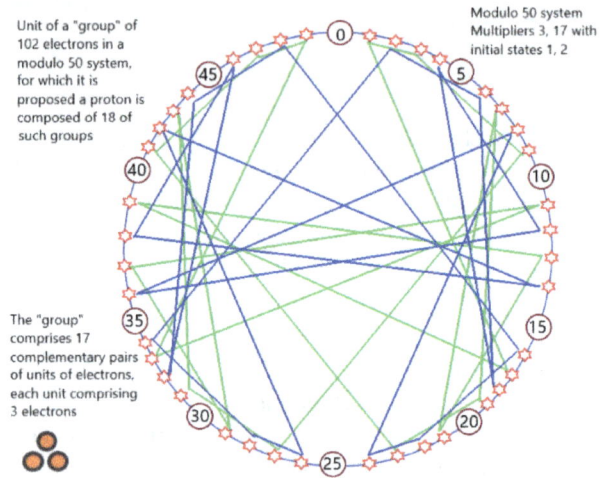

Unit of a "group" of 102 electrons in a modulo 50 system, for which it is proposed a proton is composed of 18 of such groups

Modulo 50 system Multipliers 3, 17 with initial states 1, 2

The "group" comprises 17 complementary pairs of units of electrons, each unit comprising 3 electrons

Modulo 50 circuit exhibiting 2-graphs (green, blue) of Rodin-type force graphs.

Above is diagram of two complementary Rodin-type force graphs with complementary multipliers "3" and "17". One Rodin-type force graph (green) has a start point of "1" while the other graph (blue) has a start point of "2".

In these graphs, each graph maps shifting of the respective 17 units in response to each iteration cycle through 20 possible Aether-states. Thus, each of the graphs plot the shifting of each unit over a cycle of twenty iterations. At any one time for each graph, three of the twenty available Aether-states would be vacant.

An Aether expressing Fibonacci and Lucas Sequences joined at the hip: a thought experiment

It must be first pointed out that the ultimate premise of this thought experiment is that the nature of the Aether within the nucleus of an atom is inherently different to that of the outside of the boundary which separates the nucleus from "normal"-space. Thus, while I am not aware of any observations of electrons forming triune units of "3", it is assumed for this thought experiment that any number of electrons can form units so long as they:

- support a structure supported by an MNS-F$(x)^2$ structure or an MNS-L(x) structure, or integer harmonics thereof; and they
- comprise numbers of units having values which emerge from the requirements of the Cassini Identity-type or Catalan Identity-type such that a value of "1" is achieved.

Now in this thought experiment, because of:

- each "unit" comprising three electrons;
- there being 17"-units" for each graph; and
- there being two graphs within the MNS-50 circuit,

each MNS-50 circuit supports a structure of 102 electrons. This also supposes that the triune electron units of each Rodin-type force circuit interact with a counterpart triune of the other Rodin-type force circuit.

Now the question arises, how do we arrange the eighteen MNS-50 circuits to achieve a complete "proton" of 1,836 electrons. One possible means of arranging the MNS-50

An Aether expressing Fibonacci and Lucas Sequences joined at the hip: a thought experiment

circuits to construct a proton is to allow each of the MNS-50 circuits to form themselves as triunes in the context of an MNS-14 Rodin-type circuit [= MNS-[L(3) * 2] circuit].

This is not beyond possibilities. It is already the premise that each Rodin-type force graph would have three from twenty Aether-states which are vacant. If we can assume that each MNS-50 circuit can be defined by one of the three complementary pairs of vacant states, which in turn can each be indicated as being positive or negative, then we have a possibility of each MNS-50 circuit being defined in terms of six possible Aether-value states from a total of 40 Aether-value states.

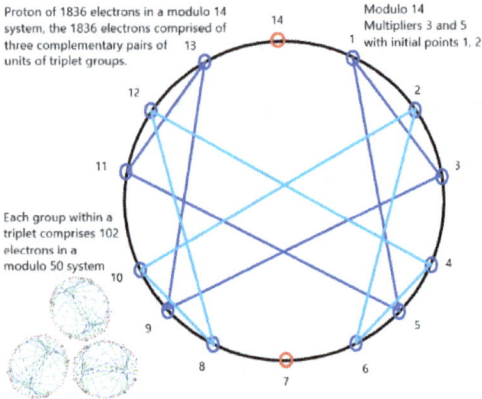

Proton of 1836 electrons in a modulo 14 system, the 1836 electrons comprised of three complementary pairs of units of triplet groups.

Modulo 14 Multipliers 3 and 5 with initial points 1, 2

Each group within a triplet comprises 102 electrons in a modulo 50 system

MNS-14 circuit comprising complementary multipliers of "3" and "5" with starting points of "1" (dark blue) and "2" (light blue) having six Aether-states each.

An Aether expressing Fibonacci and Lucas Sequences joined at the hip: a thought experiment

Such an MNS-14 circuit has:

- a complementary set of multipliers "3" and "5"; and
- two Rodin-type force graphs with start points of "1" and "2" with six Aether-states per Rodin type force graph.

If we then assume that each Rodin-type force graph supports three MNS-50 triunes, leaving three Aether-states of the MNS-14 Rodin-type force graph vacant at any one time, then we have:

- 2 [Rodin type force graphs] * 3 [MNS-50 triunes per unit] * 3 [MNS-50 circuit units per triune]

= 18 * 102 electrons per MNS-50 circuit

= 1,836 electrons.

An Aether expressing Fibonacci and Lucas Sequences joined at the hip: a thought experiment

Aethereal effects of the alternative solution as to what a proton comprises.

As postulated in Epilogue 12, the 1836 leptons **[24]** of which a proton may be considered to consist of can be broken down into 18 groups of 102 leptons. Each of the 102 lepton groups comprise two groups of 51 leptons split into 17 groups of 3 leptons which cycle through 20 states. The 20 states of one cycle of one group complement each respective state of the other cycle of the other cycle. This selection of 17 groups of 3 leptons is down to a first solution for the modified Cassini Identity of [EQN 9] identified in this book of:

$$F(x).F(-x) - (i)^2.F(x-1).F(-(x+1)) = 1 \text{ or}$$

$$F(-x).F(x) - (i)^2.F(-(x-1)).F(x+1) = 1$$

That is in the condition of X = 5 *when values are spin corrected*, then in a modulo F(x)^2 number system, then

- $5 \times 5 - (-1) \times 3 \times (-8) = 25 - 24 = 1$ or
- $5 \times 5 - (-1) \times 8 \times (-3) = 25 - 24 = 1$

To give a rough analogy of the situation, the 3 x 17 system in the Aether may be considered to be like a boat travelling on the surface of the lake, with the 22 x 8 system being the wake resulting from the motion of the 3 x 17 system in the Aether.

An Aether expressing Fibonacci and Lucas Sequences joined at the hip: a thought experiment

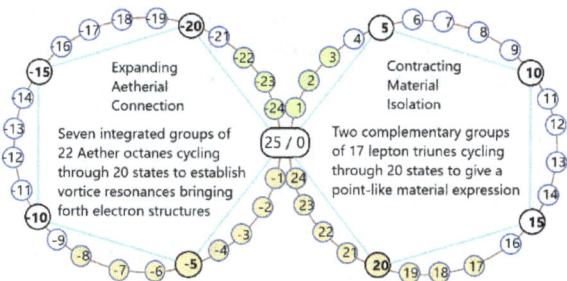

Graph illustrating view to the Aether of a modulo 25 ring number system subject to Rodin-type cycles depending upon whether the view of the ring, by the Aether, is from the left-side facing right or the right-side facing left.

However, *lets un-spin correct the values* in the two solutions of [EQN 9] in a modulo-25 system, then we obtain for the first of these solutions:

- 5 x 5 − (-1) x 3 x (25 − 8) [mod 25] =
- 25 + 3 x 17 [mod 25] =
- 25 + 51 [mod 25] =
- 0 + 1 = 1

For the second of these solutions, we obtain:

- 5 x 5 − (-1) x (25 -3) x 8 [mod 25] =
- 25 + 22 x 8 [mod 25] =
- 25 + 176 [mod 25] =
- 0 + 1 = 1

An Aether expressing Fibonacci and Lucas Sequences joined at the hip: a thought experiment

For generating a Rodin-type graph in each of the cases of the first solution and the second solution we cannot use the fundamental modulo 25 system per se.

Rather we must use number systems which provide balance for the groups cycling through 20 cycles are the first and sixth harmonics of modulo 25 [$F(5)^2$] of:

- 50 [for the group 3 x 17 of the first solution]; and
- 175 [for the group 8 x 22 of the second solution].

As illustrated before, for modulo 50 of the first solution, the following pair of 20 cycle groups in a modulo-50 environment is obtained:

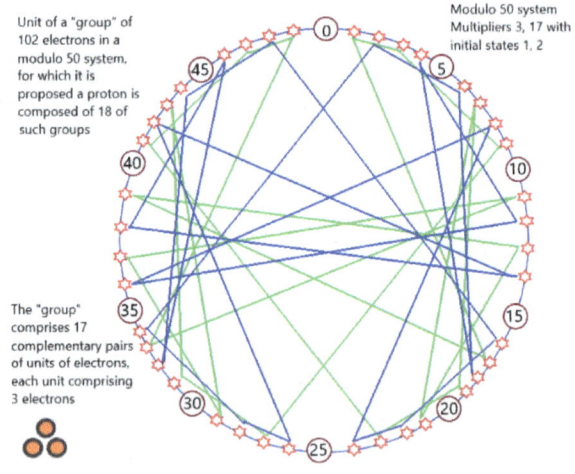

Unit of a "group" of 102 electrons in a modulo 50 system, for which it is proposed a proton is composed of 18 of such groups

Modulo 50 system Multipliers 3, 17 with initial states 1, 2

The "group" comprises 17 complementary pairs of units of electrons, each unit comprising 3 electrons

An Aether expressing Fibonacci and Lucas Sequences joined at the hip: a thought experiment

For modulo 175, the following seven 20 cycle groups in a modulo 175 environment is obtained:

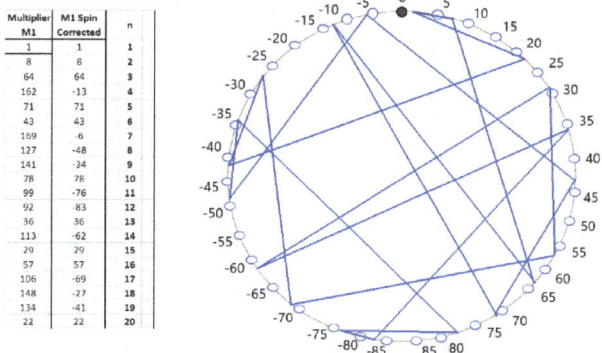

Multiplier M1	M1 Spin Corrected	n
1	1	1
8	8	2
64	64	3
162	-13	4
71	71	5
43	43	6
169	-6	7
127	-48	8
141	-34	9
78	78	10
99	-76	11
92	-83	12
36	36	13
113	-62	14
29	29	15
57	57	16
106	-69	17
148	-27	18
134	-41	19
22	22	20

Graph of modulo-175 system using multipliers 22 x 8 with initial value being equal to 1.

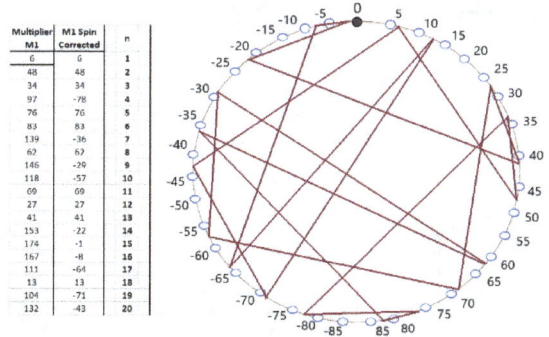

Multiplier M1	M1 Spin Corrected	n
6	6	1
48	48	2
34	34	3
97	-78	4
76	76	5
83	83	6
139	-36	7
62	62	8
146	-29	9
118	-57	10
69	69	11
27	27	12
41	41	13
153	-22	14
174	-1	15
167	-8	16
111	-64	17
13	13	18
104	-71	19
132	-43	20

Graph of modulo-175 system using multipliers 22 x 8 with initial value being equal to 6.

An Aether expressing Fibonacci and Lucas Sequences joined at the hip: a thought experiment

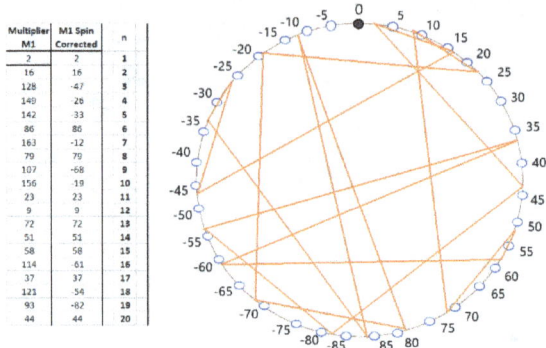

Multiplier M1	M1 Spin Corrected	n
2	2	1
16	16	2
128	-47	3
149	-26	4
142	-33	5
86	86	6
163	-12	7
79	79	8
107	-68	9
156	-19	10
23	23	11
9	9	12
72	72	13
51	51	14
58	58	15
114	-61	16
37	37	17
121	-54	18
93	-82	19
44	44	20

Rodin-type graph of modulo-175 system using multipliers 22 x 8 with initial value being equal to 2.

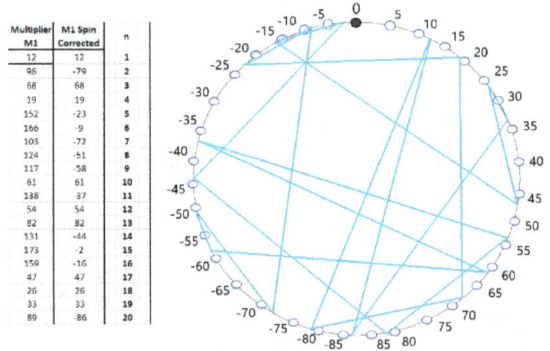

Multiplier M1	M1 Spin Corrected	n
12	12	1
96	-79	2
68	68	3
19	19	4
152	-23	5
166	-9	6
103	-72	7
124	-51	8
117	-58	9
61	61	10
138	-37	11
54	54	12
82	82	13
131	-44	14
173	-2	15
159	-16	16
47	47	17
26	26	18
33	33	19
89	-86	20

Rodin-type graph of modulo-175 system using multipliers 22 x 8 with initial value being equal to 12.

An Aether expressing Fibonacci and Lucas Sequences joined at the hip: a thought experiment

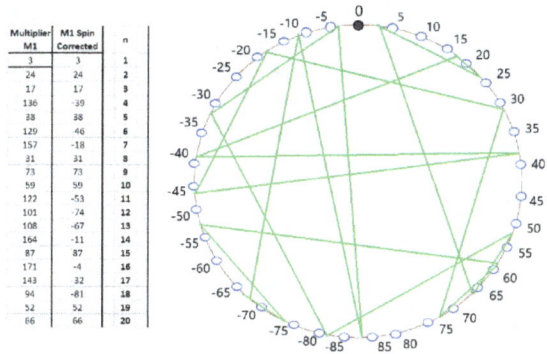

Multiplier M1	M1 Spin Corrected	n
3	3	1
24	24	2
17	17	3
136	-39	4
38	38	5
129	-46	6
157	-18	7
31	31	8
73	73	9
59	59	10
122	-53	11
101	-74	12
108	-67	13
164	-11	14
87	87	15
171	-4	16
143	32	17
94	-81	18
52	52	19
66	66	20

Rodin-type graph of modulo-175 system using multipliers 22 x 8 with initial value being equal to 3.

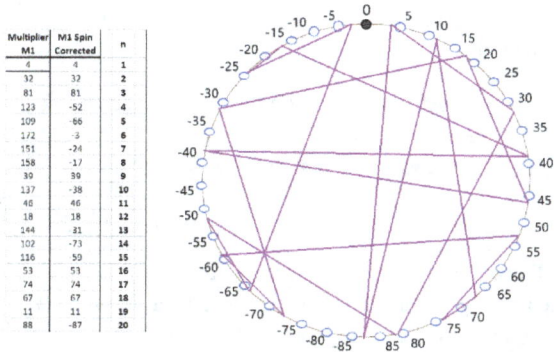

Multiplier M1	M1 Spin Corrected	n
4	4	1
32	32	2
81	81	3
123	-52	4
109	-66	5
172	-3	6
151	-24	7
158	-17	8
39	39	9
137	-38	10
48	46	11
18	18	12
144	31	13
102	-73	14
116	50	15
53	53	16
74	74	17
67	67	18
11	11	19
88	-87	20

Rodin-type graph of modulo-175 system using multipliers 22 x 8 with initial value being equal to 4.

An Aether expressing Fibonacci and Lucas Sequences joined at the hip: a thought experiment

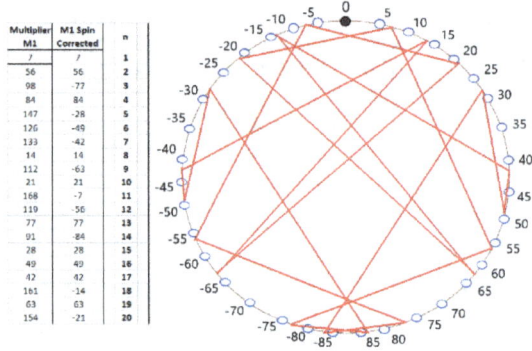

Multiplier M1	M1 Spin Corrected	n
7	7	1
56	56	2
98	-77	3
84	84	4
147	-28	5
126	-49	6
133	-42	7
14	14	8
112	-63	9
21	21	10
168	-7	11
119	-56	12
77	77	13
91	-84	14
28	28	15
49	49	16
42	42	17
161	-14	18
63	63	19
154	-21	20

Rodin-type graph of modulo-175 system using multipliers 22 x 8 with initial value being equal to 7.

Thus, for the complete description of a proton there are a total of nine cycles operating at the same time:

-     2 cycles operating on the 3 x 17 material level; and
-     7 cycles operating on the 22 x 8 aethereal level.

At this point, regarding the modulo-14 system accompanying the eighteen modulo-50 units of which the proton comprises, namely:

An Aether expressing Fibonacci and Lucas Sequences joined at the hip: a thought experiment

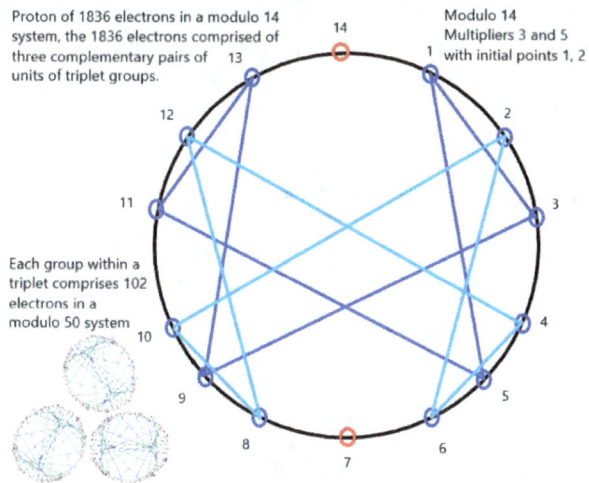

Proton of 1836 electrons in a modulo 14 system, the 1836 electrons comprised of three complementary pairs of 13 units of triplet groups.

Modulo 14
Multipliers 3 and 5
with initial points 1, 2

Each group within a triplet comprises 102 electrons in a modulo 50 system

Rodin-type graph of modulo-14 system using two groups of multipliers 3 x 5 with one group having an initial value of 1 and the second group having an initial value of 2.

It appears that the modulo 14 system results from an interaction between

- the initial "2" cycles of the 3 x 17 solution and
- the complementary "7" cycles of the 22 x 8 solution.

An Aether expressing Fibonacci and Lucas Sequences joined at the hip: a thought experiment

Epilogue 14:

Analysis of Material and Aethereal structures of the proton in terms of modulo system syntax the Cassini and Catalan Identities.

In Epilogue 13 we explored the possibility of using

- a combination of 18 modulo-25 systems, or at least the 1st harmonic thereof (modulo-50),

in combination with

- a modulo-7 system, or at least the 1st harmonic thereof (modulo 14)

to describe a proton in terms of a model comprising 1836 leptons based on the proton to electron mass ratio. This is based on one solution of the modified Cassini Identity in which a combination 17 groups of 3 leptons are used about an inversion plane. We have not yet however discussed the second solution of the modified Cassini Identity of 22 x 8. In this case of the second solution, we will not discuss the relationship in terms of mass bearing leptons but in terms of, for a want of a better word, "aethereons" which do not carry mass. Rather it is proposed that these "aethereons" of the proton interact with the Aether on the shadow side of the nucleus boundary. It is proposed that this interaction on the shadow side seeks to communicate structure from the proton(s), through resonance, to the aether when the aether about the nucleus is in the presence of a suitable numbers of leptons in the form of electrons.

An Aether expressing Fibonacci and Lucas Sequences joined at the hip: a thought experiment

The following is an analysis of the properties of the standing and dynamic aethereal effects resulting from proposed syntax of modulo-25 systems. In addition, there follows a breakdown analysis of the syntax outcomes of the modulo-25 system with possible Aether vector resonances in view of the Cassini and Catalan identities.

Aethereal Environment of the Proton in terms of "words" and "carrier states" of modulo-25

Regardless of whether dynamic cycles of "3" x "17" or "22" x "8" are operating with the modulo-25 number system, the words generated by the modulo-25 system per se will still consist of 4-bits length between punctuations of resulting from multiples of "5". Driving of the Fibonacci sequences through the Aether-based modulo 25 number system, the words of the modulo-25 will still give rise to:

-   a "carrier state" word of a standing aether value of "-5" or "20" (or also "-5" when spin-corrected); and
-   Supra-MNS states in sequence of "1", "18", "24" and "7" derived from a dynamic cycle of letters of Aether states "-7" x "7" being multiplied along the inversion line with an initial value of "1".

In addition, the modulo-25 number system holding each of the "3" x "17" and "22" x "8" inversion plane cycles also give rise to static values of

An Aether expressing Fibonacci and Lucas Sequences joined at the hip: a thought experiment

- "1" and "2" for respective ones of the 3 x 17 cycles; and
- "1", "2", "3", "4", "6", "7" and "12" for respective ones of the 22 x 8 cycles .

Factorising "14" of the modulo-14 system

As noted, the value "14" of the modulo-14 system can be factored into "2" x "7". This may be down to the number of cycles in the 3 x 17 and 22 x 8 being 2 and 7 respectively, and possibly positively reinforced by the static values "2" and "7" generated by the inversion-plane cycles. It is noted that values "2" and "7"are each a respective value of the Lucas system. That is the static value "2" is L(0) and the static value "7" is L(4). In a system based on L(0) x L(4) then we see that this combination of multipliers is one half of a Catalan Identity where:

- $L(2)^2 - L(0) \times L(4) = -5$

As such the only element missing here, which has not been generated anywhere else is $L(2)^2$ which is equal the difference between values which are already available:

- $L(2)^2 = L(0) \times L(4) - 5 = 9.$

But as it turns out $L(2)^2 = F(4)^2$. So, in this respect this model of the Proton also has at its centre a virtual static Aether value of $F(4)^2$.

320

An Aether expressing Fibonacci and Lucas Sequences joined at the hip: a thought experiment

In this respect I am stealing an idea from semiconductor physics. Namely one can have a charge carrier of an electron where there is a superfluous electron in a silicon matrix due to the addition of an n-type impurity in the silicon matrix. Conversely one can have a "hole" which may also act as a charge carrier due to the addition of an impurity, but in which the impurity is p-type and lacks an electron compared to each silicon atom in the matrix. In this case the virtual $F(4)^2$ or $L(2)^2$ is the "hole" which provides a shadow version of the virtual $F(4)^2$ or $L(2)^2$.

Important to note, it is the proposition here that this shadow version also manifests on the other side of the boundary which separates the nucleus from the outer Aether.

The story however does not end here. The generation of *the virtual or shadow $F(4)^2$* vortex centre is also considered to be dependent on the presence of a "-5" static Aether state from the modulo-25 number system. That is the modulo-25 number system holds each of the complementary inversion plane cycles of

- the constricting 3 x 17 cycle and
- the extending 22 x 8 cycle

at the boundary of a vortex formed at either

- the centre of a circle or
- one of two focus points of an ellipse

321

An Aether expressing Fibonacci and Lucas Sequences joined at the hip: a thought experiment

defined by the modulo-25 system, and this modulo-25 system generates carrier state words of either "-5" and the unspin corrected value of "20". In terms of a modulo-$F(4)^2$ system however, considering the generation of "-5" carrier state in the modulo-25 system, this value of "-5" is also:

- the unspin-corrected value of the carrier state value "4" of the modulo-$F(4)^2$ (=9) modulo system; and
- an inverse polarity of one of the multipliers of the inverse plane as expressed in the Cassini Identity.

In addition, it is further noted that the Supra-MNS values of "7" and "7" of the modulo-25 system, in the modulo-9 system, correspond to a value of "-2". It is submitted here therefor that this combination of Identities, "supra-MNS states" (information states) and "carrier states" (i.e., analogous to a carrier signal in radio transmission) gives rise to a shadow modulo-9 system beyond the boundary of the nucleus holding proton. Furthermore, it is submitted that this shadow modulo-9 system is waiting for leptons, in the form of electrons, to interact with it to create a complementary modulo-9 system of five paired electrons. This may occur for example because of 10 protons being in a nucleus, as in the case of neon, giving rise to a requirement for ten electrons which are induced to resonate with the shadow modulo-9 number system of aethereons.

An Aether expressing Fibonacci and Lucas Sequences joined at the hip: a thought experiment

A relationship between "176" aethereons of the modulo-175 system and "51" leptons of the modulo-50 system

As mentioned earlier, the 1836 leptons can be split into eighteen complementary pairs of 3 x 17 – 20-state cycles. The combination of complementary "3" and "17" of the eighteen complementary pairs of the modified Cassini Identity however, being of prime numbers, cannot be factorised any smaller. It is noted however that aether-state "17" could be equal to

- 34 / 2 =
- F(9) / F(3).

If these values of F(3) and F(9) were to relate to a Catalan Identity involving "34" and "2" as operators, then this would involve F(6) x F(-6) at the centre, or 8 x -8 = -64. This modified Catalan Identity of:

- F(9) x F(3) – F(6) x F(-6)

gives rise to a value of "4". This aether value of "4" can in turn can lead us to "4" being either:

- the aether "carrier state" of either the closed modulo number system of "4" [F(3) x F(-3)] or the closed modulo number system of "-9" [F3(4) x F(-4)]; or
- the square of the "2" operator L(0), i.e., L(0)^2.

An Aether expressing Fibonacci and Lucas Sequences joined at the hip: a thought experiment

Regardless, I'm going on a rabbit trail here and let's get back to the discussion at hand.

It is assumed that the mass of a lepton, of which the proton is considered to be composed of 1836 thereof, is the same as the mass of an electron which is external to the nucleus about the proton. In contrast we have no idea if aethereons employed in the seven 22 x 8 20-cycle states have any "mass" oar any equivalent thereof. Nor can we *prima facie* expect that there are only as many units of 22 x 8 aethereons as there are units of pairs of 3 x 17 leptons.

At this point, it appears pertinent to try to describe in terms of an analogy with different phases of water of liquid, ice, and water vapour / steam, to explain the relationship between leptons, the aether, and aethereons. In these cases, the aether can be described as a liquid phase, leptons as an ice phase, and aethereons as a water vapour / steam phase.

We do know however that each intermediate lepton group of complementary pairs of 3 x 17 leptons may suppose the presence of 102 leptons (and that there are eighteen of these groups). We then suppose that each group of aethereons comprise 22 x 8 times 7 (i.e., for the number of different cycles which may be implemented). If we also accept the number of groups aethereons is the same as the number of groups of leptons (that is, eighteen), then in total we should have:

- 22 x 8 x 7 x 18 =
- 22,176 aethereons

An Aether expressing Fibonacci and Lucas Sequences joined at the hip: a thought experiment

or

- 126 times groups of 176 aethereons.

Of course, when dealing with a value of "126" as the number of "176"-units, we are dealing with a set of inversion plane multipliers of either:

- "9" $[F(4)^2]$ x "14[L(0) x L(4)]", or
- "6" [F(3) x F(4)] x "21" [F(8) or L(2) x L(4)]

in a closed modulo-125 [= $F(5)^3$]number system, the aether value "125" being the fourth harmonic of a closed modulo-25 number system as well as being the carrier state of a closed modulo-$F(7)^2$ [or 169] system.

In terms of a connection between the 3 x 17 cycles and the 22 x 8 cycles, we see a connection manifesting through use of at least:

- the fourth iteration of the Fibonacci sequence or the second iteration of the Lucas sequence [F(4) = 3 = L(2)] and
- the "2" operator of L(0) [or F(3)].

One must also note that "7" and "18" are themselves drive multipliers used to produces Supra MNS values ("information states)" in any harmonic of the closed modulo 25 number system. Once again, we can see that there is mutual support among all aetheric and structural outcomes within the proton structure.

An Aether expressing Fibonacci and Lucas Sequences joined at the hip: a thought experiment

Factorising the "176" aethereons of the modulo-175 system

The number "176" can be factorised for obvious reasons into:

- "22" x "8"

and these numbers can be further broken down into

- "2" x "11" and "2" x "4"; or
- "2" x "2" x "11" x "4"; or
- "4" x "44"

If we try to express these values in terms of Lucas numbers, we gat values of either:

- $L(0) \times L(0) \times L(5) \times L(3)$; or
- $L(0)^2 \times L(5) \times L(3)$; or
- $L(3) \times L(3) \times L(5)$; or
- $L(3)^2 \times L(5)$.

In dealing with the aether state "44" generated by Lucas aether values $L(5) \times L(3)$, this is interesting in terms of two aspects:

- the aether-value "44" is equal to the modulus of the aether-value "-44" of the aether carrier states of each of a closed modulo-$F(6)^2$ system [modulo-64] and modulo-$F(7)^2$ system [modulo-169]; and
- the relationship $L(5) \times L(3)$ is one half of a Cassini Identity involving Lucas values, that is:
  - $L(-3) \times L(5)) - L(-4) \times L(4) = -5$; or

An Aether expressing Fibonacci and Lucas Sequences joined at the hip: a thought experiment

- L(3) x L(-5) – L(4) x L(-4) = -5.

In latter case, even if we take the case that the Cassini Identity applied to the aether state "44" gives rise to an aether value of modulus "5" normal thereto, then the resulting product of "4" x "5" = "20". Looking back to the carrier state of a modulo-25, the aether value "20" is the alternate to the spin-corrected value of "-5".

In turn, if we turn back to Modulo-9 systems, and the Cassini Identity associated there with in which there are driving multipliers of "5" and "2",then:

- the aether state "20" in context of a closed modulo-9 system [modulo-F(4)^2] becomes an aether value "2" ["20" – "18" = 2]which is one of the inverse plane multipliers "2" and "5";
- the factors "4" and "5" of "20" are in turn the main aether carrier state value and the other of the inverse plane multipliers "2" and "5" of the closed modulo-9 system.

Thus, again we see various interconnected and mutually supporting relationships between aether states involving:

- the closed modulo-25 number system and its harmonics; and

327

An Aether expressing Fibonacci and Lucas Sequences joined at the hip: a thought experiment

- the supra-MNS state and carrier state words generated by the fundamental modulo-25 system,

whereby they are interconnected with:

- one or other of the carrier states of closed modulo-64 [modulo-F(6)^2] and modulo-169 [modulo-F(7)^2]
- both aether carrier-state words of the modulo-9 [modulo-F(4)^2] as well as the drive multipliers of the pair of inverse plane multipliers giving rise to the supra-MNS aether state words "1", "2", "4", "8", "7" and "5" of the closed modulo-number system.

Ultimately, it is submitted that it is through these interplay of number relationships of:

- the "22" x "8" aethereons operating under closed modulo-175 systems outside of the nucleus within which a proton is held,

linked to

- the "3" x "17" leptons operating under closed modulo-50 systems internal to the nucleus,

which enable structures to be created between atoms. In these cases, it is submitted that the enabling of structures of groups of atoms giving rise to material form of existence

328

An Aether expressing Fibonacci and Lucas Sequences joined at the hip: a thought experiment

are at least partly achieved through the generation of the "aethereons". It is submitted that interactions between leptons external to a nucleus (i.e., electrons) work to create material structures. Further, it is submitted that

- those structures that are enabled through interactions between electrons and aethereons are those structures which resonate with closed modulo-$f(n)^2$ systems and harmonics thereof,

in together with

- limitations imposed by the multipliers of the Cassini and Catalan identities meeting the requirements of the inversion plane. That is in each closed modulo system of value "x":
  - $F(y) \times F(z) = 1$ [modulo-x].

Units of "3" in a modulo-5 [F(5)] system and "22" in a modulo-21 [L(4) x L(2)] system

Regarding the unit of "3" of the "3" x "17" cycle system, each unit of "3" too must follow a single cycle of states.

An Aether expressing Fibonacci and Lucas Sequences joined at the hip: a thought experiment

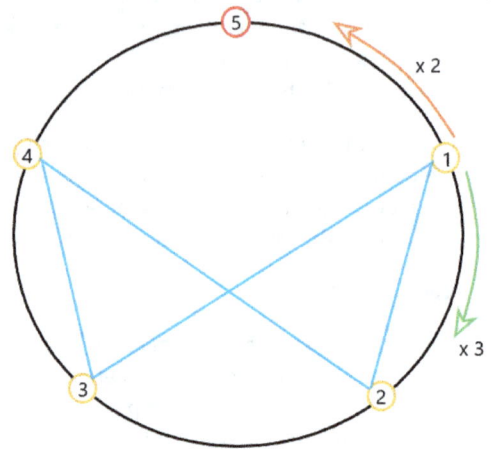

Rodin-type force graph for Modulo-5 with multipliers

-      **"2" and "3"**

This single cycle of four states also imitates the first four states of the Lucas sequence from "L(0)" to "L(3)", cycling through "2", "1", "3" and "4" in a forward direction.

As it stands, as a modulo-5 system using "3" as a multiplier (for the number of leptons), the operator "2" acts as a complementary inversion plane multiplier. In addition, in common with each of the inversion plane cycles involving "3" x "17" and "22" x "8", the number of states in a Fibonacci number cycle subject to modulo-5 is also twenty.

An Aether expressing Fibonacci and Lucas Sequences joined at the hip: a thought experiment

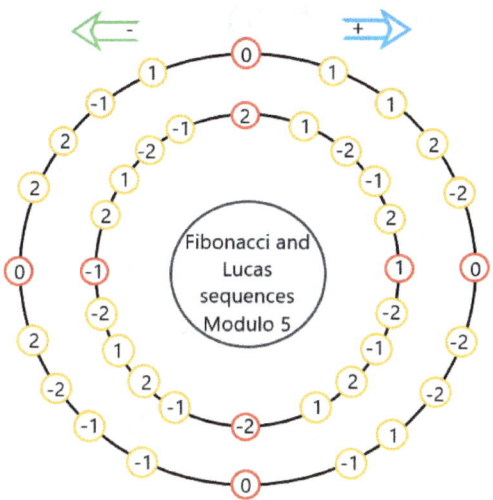

20 Aether-elements with modulo-5 of each of the Fibonacci sequence and the Lucas sequence

In contrast regarding the unit "22" of the "22" x "8" of the cycle system, this does not have a singular cycle of states. Rather, with modular-21, there are a possible two paired-cycles and a single oscillatory cycle.

An Aether expressing Fibonacci and Lucas Sequences joined at the hip: a thought experiment

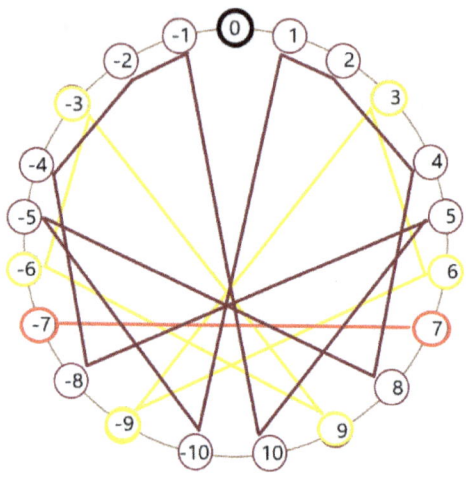

Modulo-21 cycles with multipliers of

- **"2" [L(0)] and "11" [L(5)]**

It is submitted however that there is no single solution as to how the twenty-two / eleven pairs of aethereons within the closed modulo-21 system may be distributed.

If we look at the cycles with initial aether values of "1" and "-1", then we have paired cycles:

An Aether expressing Fibonacci and Lucas Sequences joined at the hip: a thought experiment

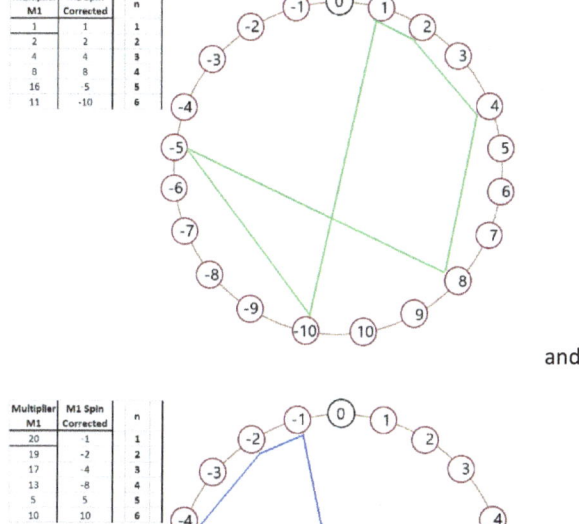

and

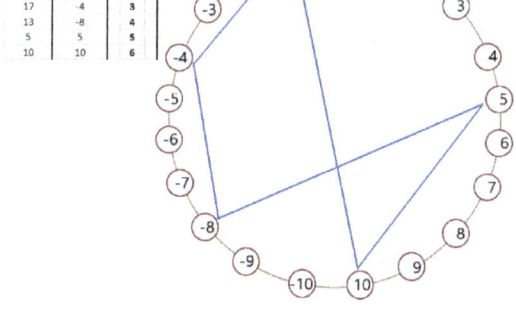

Complementary six state cycles

Assuming at least one vacancy for each cycle of six, and a maximum of three vacancies, the number of pairs of aethereons held by this pair of complementary cycles may

An Aether expressing Fibonacci and Lucas Sequences joined at the hip: a thought experiment

vary between six (three pairs per cycle and three vacancies) or ten (five pairs per cycle and one vacancy).

If we look at the cycles with initial aether values of "3" and "-3", then we have paired cycles:

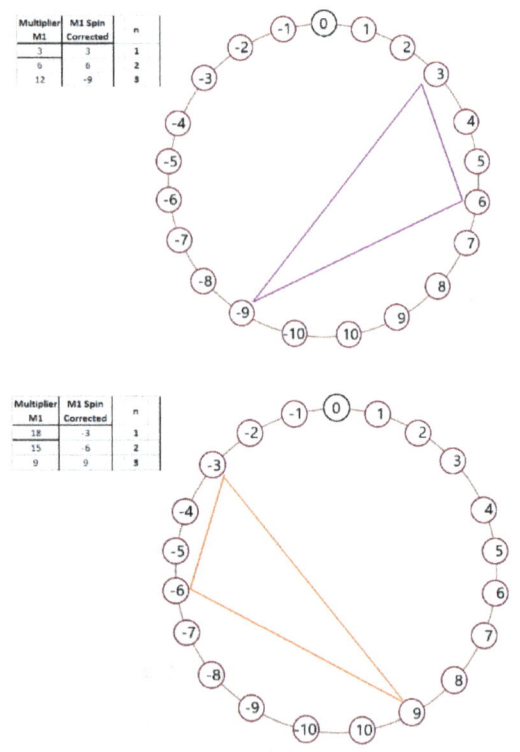

Complementary three state cycles

An Aether expressing Fibonacci and Lucas Sequences joined at the hip: a thought experiment

Assuming at least one vacancy for each of these cycles involving only three state, and a maximum of three vacancies, the number of pairs of aethereons held by these pairs of complementary cycles may vary between

- four (two pairs per cycle and one vacancy) or
- zero.

Finally for the last cycle acting as a metronome for the other two pairs, we have an initial value of "7" or "-7" defining a cycle of a single pair of aethereons.

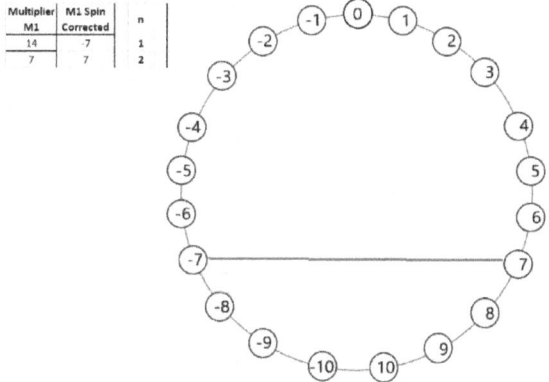

Multiplier M1	M1 Spin Corrected	n
14	7	1
7	7	2

Oscillatory two state cycle having metronome function.

An Aether expressing Fibonacci and Lucas Sequences joined at the hip: a thought experiment

Dependence of material phases determined by arrangements of closed modulo-21 cycles.

As mentioned earlier, there appears to be no single solution in which to arrange the modulo-21 cycles involving 11 pairs of aethereons. It is submitted however that the solution to determining which arrangement is undertaken is dependent on an outside agency with properties analogous to available thermal energy in the form of heat, i.e. subject to temperature and pressure.

Cycle Type / phase	6-state cycle pairs	3-state cycle pairs	1-state cycle pairs
1	5	0	1
2	4	1	1
3	3	2	1

Table for distribution of pairs to cycle per phase. Total number of pairs in each case add up to

- **5 x 2 + 1 = 11 pairs of aethereons or**
- **22 aethereons.**

It is noted however that even if this model for the generation of phases of materials (i.e., for phases of gas, liquid, solid) holds, it can nevertheless only be speculated as to whether e.g., "phase 1" corresponds to solid or gas and "phase 3" corresponds to gas or solid.

An Aether expressing Fibonacci and Lucas Sequences joined at the hip: a thought experiment

Aethereons and the Fine Structure Constant.

To cite from the definition of the Fine Structure Constant [25] in Wikipedia (RTM):

> "In physics, the fine-structure constant, also known as the Sommerfeld constant, commonly denoted by α (the Greek letter alpha), is a fundamental physical constant which quantifies the strength of the electromagnetic interaction between elementary charged particles.
>
> It is a dimensionless quantity, independent of the system of units used, which is related to the strength of the coupling of an elementary charge e with the electromagnetic field, by the formula $4\pi\varepsilon_0\hbar c\alpha = e^2$. Its numerical value is approximately 0.0072973525693 ≃
>
> 1 / 137.035999084,
>
> with a relative uncertainty of 1.5×10−10."

-   "Fine Structure Constant", Wikipedia (RTM) as of 26.08.2023

I may be clutching at straws at this point, but the value "137" per se is also equal to

-   $L(5)^2 + L(3)^2 = 121 + 16 = 137$.

In the case of the closed modulo-21 cycle we have two drive multipliers of L(5) operating in tandem with one

An Aether expressing Fibonacci and Lucas Sequences joined at the hip: a thought experiment

another to drive the paired 6-state cycles, paired 3-state cycles, and metronome cycle. It is submitted that if even the inversion plane L(5) driver multipliers of the paired cycles can be considered to work in tandem with one another, then the result would be a force operating normal to the aethereons of $L(5)^2$.

In addition, the value "21" of the closed modulo-21 system is itself a product of L(4) x L(2) [i.e., "7" x "3"]. This product can itself be considered one half of a Cassini relationship of:

-    L(3) x L(-3) + L(4) x L(-2) = -16 + 21 = 5 or
-    L(-3) x L(3) + L(-4) x L(2) = -16 + 21 = 5

in an environment in which the "5" is already present due to the carrier state of the closed modulo-175 system with the "22" x "8" inversion plane drivers. In this respect it is submitted the environment of the modulo-21 system in combination with the closed modulo-175 system of which it is apart, gives rise to a virtual $L(3)^2$ result in which:

-    L(4) x L(-2) - 5 = - L(3) x L(-3) = 16 or
-    L(-4) x L(2) - 5 = - L(-3) x L(3) = 16

wherein the sum of

-    the $L(5)^2$ aether value of the two L(5) multipliers operating in tandem [= 121], and
-    the $L(3)^2$ aether value operating normally to the closed modulo-21 system [= 16],

gives rise to a value of "137". But as mentioned earlier, I may be clutching at straws for this one.

An Aether expressing Fibonacci and Lucas Sequences joined at the hip: a thought experiment

Conclusion

We may or may not live in a holographic universe with an aether acting as a framework. But if we do, and I have convinced You of this as a possibility, then it is suggested that mathematic properties of

- algebraic sequences such as Fibonacci and Lucas Sequences, and
- Identities derived from these sequences such as the Cassini Identity and Catalan Identity

be studied in order to better understand the limitations and possibilities of this holographic universe.

If You have read to the end, thank you for your patience and curiosity. If my writing has not in any way inspired You, I hope at least it has given You some food for thought.

An Aether expressing Fibonacci and Lucas Sequences joined at the hip: a thought experiment

References:

[1] Fibonacci | Biography, Sequence, & Facts | Britannica

[2] Lucas number - Wikipedia

[3] Generalizations of Fibonacci numbers - Wikipedia

[4] Lucas sequence - Wikipedia

[5] Complex number - Wikipedia

[6] Quaternion - Wikipedia

[7] Dual number - Wikipedia

[8] Cassini and Catalan identities - Wikipedia

[9] Giovanni Domenico Cassini - Wikipedia

[10] Johannes Kepler - Wikipedia

[11] LC circuit - Wikipedia

[12] Mandelbrot set - Wikipedia

[13] What It's Like to Actually See an Atomic Explosion | RealClearScience

[14] Spin-1/2 - Wikipedia

[15] Kepler's laws of planetary motion - Wikipedia

[16] Microsoft Word - Rodin - The Rodin Number Map and the Rodin Coil (rexresearch.com)

An Aether expressing Fibonacci and Lucas Sequences joined at the hip: a thought experiment

[17] Marko RODIN -- Collected Papers & Videos (rexresearch.com)

[18] Moon - Wikipedia

[19] Book of Enoch - Wikipedia

[20] Chapter 71 [SECT. XIII.] – The Book of Enoch (bookofenoch.com)

[21] Sidereal time - Wikipedia

[22] Solar rotation - Wikipedia

[23] Proton-to-electron mass ratio - Wikipedia

[24] Lepton - Wikipedia

[25] Fine-structure constant - Wikipedia